P9-CBK-332

Making Nature, Shaping Culture

Plant Biodiversity in Global Context

VOLUME 8 IN THE SERIES
Our Sustainable Future

Making Nature Shaping Culture

Plant Biodiversity in Global Context

Lawrence Busch, William B. Lacy, Jeffrey Burkhardt,
Douglas Hemken, Jubel Moraga-Rojel,
Timothy Koponen, and José de Souza Silva

University of Nebraska, Lincoln and London

Manufactured in the United States of America.
The paper in this book meets the minimum requirements of American National Standard for Information Sciences – Permanence of Paper for Printed Library Materials, ANSI Z39.48-1984.

Illustrations on pp. 1, 69, 129, and 183 are details from John Ellis, *Directions for Bringing over Seeds and Plants*, 1770. Courtesy Library, the Academy of Natural Sciences of Philadelphia.

Library of Congress Cataloging-in-Publication Data. Making nature, shaping culture : plant biodiversity in global context / Lawrence Busch . . . [et al.]. p. cm. (Our sustainable future) Includes bibliographical references (p.) and index. ISBN 0-8032-1256-9 (cl : alk. paper)
1. Food crops — Germplasm resources.
2. Germplasm resources, Plant.
I. Busch, Lawrence.
II. Series. SB123.3.M34 1996
306.3′49 — dc20
95-10244 CIP

To our spouses, parents, siblings, and children and
to the diversity of our cultures.

Contents

Tables

Abbreviations

ABRABI	Associção Brasiliera de Empresas da Biotechnologia
ABRASEM	Brazilian Seed Association
AES	Agricultural Experiment Station
AID	Agency for International Development
AIES	Association Internationale d'Essais de Semences (France)
ALHFAM	Association of Living Historical Farms and Agricultural Museums
ARS	Agricultural Research Service
ASTA	American Seed Trade Association
BAG	Active Banks of Germplasm
BRG	Bureau de Ressources Génétiques (France)
CAST	Council for Agricultural Science and Technology
CENARGEN	National Center for Genetic Resources and Biotechnology
CET	Centro de Education y Technologia (Chile)
CGIAR	Consultative Group for International Agricultural Research
CIAT	Centro Internacional de Agricultura Tropical (International Center for Tropical Agriculture)
CIB	Interministerial Council for Biotechnology
CIMMYT	International Maize and Wheat Center
CIP	Centro International de la Papa (International Potato Center)
CIRAD	Centre de Coopération Internationale en Recherche Agronomique pour le Développement
CNPq	National Research Council for Scientific and Technological Development
CNPSO	National Soybean Research Center

CNRS	Centre Nationale de la Recherche Scientifique (France)
CONAF	National Forestry Corporation (Corporación Nacional Forestal [Chile])
CONICYT	National Council for Science and Technology
COV	Certificat d'Obtention Végétale (France)
CPGR	Commission on Plant Genetic Resources
CPOV	Comité de la Protection des Obtentions Végétales (France)
CSRS	Cooperative State Research Service
CTPS	Comité Technique Permanent de la Sélection des Plantes Cultivées (France)
DARPA	Defense Advanced Research Projects Agency
DHS	distinctness, homogeneity, and stability
DNPEA	Departamento Nacional de Pesquisa e Experimentacão Agricola (Brazil)
DOV	Droit d'Obtention Végétale (France)
EMBRAPA	Empresa Brasileira de Pesquisa Agropecuaria (Brazilian Public Enterprise for Agricultural Research)
FAO	Food and Agriculture Organization (UN)
FDA	Food and Drug Administration
GAO	General Accounting Office
GATT	General Agreement on Tariffs and Trade
GEF	Global Environment Facility
GELI	Group d'Etudes des Lignées (France)
GEVES	Group d'Etudes et de Côntrole des Variétés et des Semences (France)
GNISP	Groupement National Interprofessionel de Semences des Plantes (France)
GRAIN	Genetic Resources Action International
GRIN	Germplasm Resources Information Network
HYV	High-yielding varieties
IAC	Instituto Agronomico de Campinas (Brazil)
IARC	International Agricultural Research Center
IBP	International Biological Programme
IBPGR	International Board for Plant Genetic Resources
ICARDA	International Center for Agricultural Research in the Dry Areas
ICRISAT	International Crops Research Institute for the Semi-Arid Tropics

ICSV	International Council of Scientific Unions
IDRC	International Development Research Centre
INIA	National Agricultural Research Institute (Chile)
INPI	National Industrial Property Institute
INRA	Institut Nationale de la Recherche Agronomique (France)
IOCU	International Organization of Consumers Unions
IPB	International Plant Breeders
IPGRI	International Plant Genetic Resources Institute
IPND	Primeiro Plano Nacional de Desenvolvimento (Brazil)
IRRI	International Rice Research Institute
IRTP	International Rice Testing Program
JICA	Japanese International Cooperation Agency
KNPS	Kentucky Native Plant Society
LHF	Living Historical Farm
NAFEX	North American Fruit Explorers
NAS	National Academy of Sciences
NASA	National Aeronautic and Space Administration
NBS	National Biological Survey
NGO	Nongovernmental Organization
NGRL	National Germplasm Resources Laboratory
NGRP	National Genetic Resources Program
NPGS	National Plant Germplasm System (U.S.)
NRC	National Research Council
NRSP	National Research Support Project
NS/S	Native Seeds/SEARCH
NS	Native Seeds
NSSL	National Seed Storage Laboratory
OECD	Organisation for Economic Cooperation and Development
ONIC	Office National Interprofessionel des Céréales (France)
ORSTOM	Institut Français de Recherche Scientifique pour le Développement en Coopération (France)
OTA	Office of Technology Assessment
PBR	Plant Breeders' Rights
PCR	Polymerase chain reactions
PGRP	Plant Genetic Resources Project
PNPB	National Research Program in Biotechnology
PNRG	National Program for Genetic Resources

PPA	Plant Patent Act
PVP	Plant Variety Protection
PVPA	Plant Variety Protection Act
RDA	Recommended daily allowance
RFLP	Restriction fragment length polymorphisms
RRC	Rodale Research Center
SAES	state agricultural experiment stations
SAG	Agricultural and Livestock Service
SBPC	Sociedade Brasileira para o Progresso da Ciencia (Brazilian Society for the Progress of Science)
SBRC	Brazilian System for the Registry of Cultivars
SCPA	Brazilian Cooperative System of Agricultural Research
SEARCH	Southwestern Endangered Arid-Land Resource Clearing House
SNA	Sociedad Nacional de Agricultura (Chile)
SOC	Service Officiel de Côntrole et de Certification (France)
SOLAGRAL	Solidarités Agro-Alimentaires
SSE	Seed Savers Exchange
UACH	Universidad Austral de Chile (Chile)
UAPB	Support Unit for Research in Biotechnology
UNCED	United Nations Conference on Environment and Development
UNEP	United Nations Environment Programme
UPOV	Union pour la Protection des Obtentions Végétales
USDA	United States Department of Agriculture
WIPO	World Intellectual Property Organization
ZPC	Cooperative Association for Improved Seeds of Holland

Acknowledgments

The authors wish to thank Fred Buttel, Laura Lacy, Mark Sagoff, and the reviewers for the University of Nebraska Press for their helpful comments on parts of a previous draft of this manuscript. We would also like to thank Keiko Tanaka for her help in compiling the list of abbreviations. This volume was written with support from the National Science Foundation's Ethics and Values Studies Program (EVS) and the Cooperative State Research Service (USDA) under grant No. BSR-8608719 as well as the Michigan and Pennsylvania Agricultural Experiment Stations. The chapter on France was written in large part while the senior author was a visiting scientist (*directeur de recherches*) at ORSTOM in Paris. Opinions, findings, and conclusions are solely those of the authors and do not necessarily reflect the views of the sponsors.

Show me your garden and
I will tell you what you are.

Alfred Austin, *The Garden that I Love*

PART ONE

Worlds in Collision

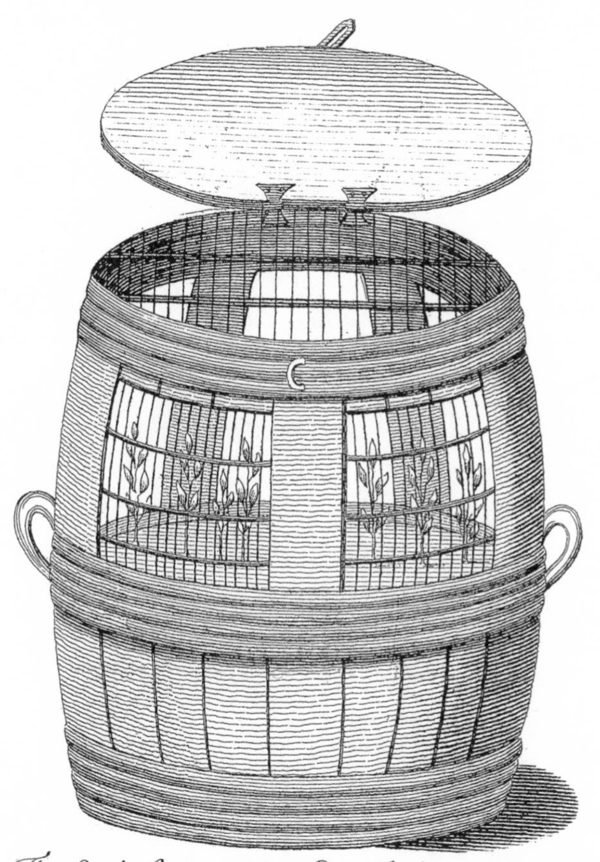

The Cask for sowing East India seeds with the openings defended by Wire.

I

Introduction: Constituting Nature, Manufacturing Plants

The moral is that every tree needs labour.
Virgil, *Georgics*

Nature is not natural. The idea that there is something "out there" – wilderness, the cosmos, physical or biological "reality" – that sciences discover, analyze, map, and manipulate is as mistaken as the idea that sculptors have occasionally espoused that the statue is already in the piece of stone, only to be revealed by the carving. In truth, the statue exists only through the carving. Similarly, nature exists only through its description, analysis, mapping, and manipulation. This is not to say that things lack existence independent from humans. But the idea that the nature we know – the collection, array, or system of those things – is independent of human intention, evaluation, and action rests on two fundamental errors, one historical and anthropological, the other epistemological. A central argument of this book is that these errors need to be recognized and corrected, conceptually and politically.

We have populated this planet for millennia. We have cultivated the earth, even in the very process of not cultivating it. Hunters and gatherers, for example, selected their food plants and animal prey. They appropriated things from their environments in a variety of ways. They consumed and used those plant and animal products. But they also learned to identify, classify, and remember the vine, weed, paw-print, and the like, appropriate to their needs and wants. Even before the dawn of agriculture and civilization, science and art, nature was already the sum of the useful and useless, edible and inedible, fearless and fearful, desired and despised, loved and hated. Nature was what human beings made it, for their ends.

There are also epistemological reasons for arguing that nature is not natural or independent of human involvement. Although philosophers and scientists have tried for centuries to distinguish what is essential (i.e., objective and inherent) to things from what is phenomenal (i.e., subjective and variable), philosophers now acknowledge that the distinction itself may be misguided. Taking a clue from physicists such as Werner Heisenberg, who maintained that the very process of observation affects and reconstitutes the phenomenon under investigation, many and perhaps even most epistemologists now hold that what we call objective reality is constituted by both the actual physical configurations of the elements in things and human conceptual frameworks (theories, definitions, and "facts"). Nature may be, as the seventeenth-century philosopher Baruch Spinoza defined it, "All there is . . . (that is conceived by God)," but nature as we know it is, indeed, nature as we know it. It is our ordering of the information received by our senses that constitutes the picture of "all that is" and that we refer to as nature. We may not create the molecules, organisms, or systems among organisms, but we nevertheless constitute nature through our practical and cognitive activities.

How we create nature is worthy of further discussion, for as we shall show, it is not merely an arcane matter of interest only to philosophers but a practical matter that should be of interest to all. It is, for example, both mistaken and arrogant to say that individuals constitute nature *de novo* and by themselves. People are born and raised in families and in cultures. People learn concepts from religion, from formal and informal schooling, from political institutions, just as they learn techniques and technologies for appropriating nature. People's practices, concepts, theories, and values with respect to nature are essentially social in origin. Although individuals may occasionally have different, even revolutionary, notions of what is and how that "what" works, even revolutionary ideas start from the conceptual framework provided by a community or society (see Elster 1983). To put it differently, nature is socially constituted. There is considerable agreement among societies as to what nature and natural processes are, in part as a result of commerce and communication, yet societies differ.

A crucial point for our analysis is how societies or communities constitute nature. Nature has been and continues to be constituted in an "unnatural" way in most industrialized nations. Contemporary science is a particularly interesting case. In the very process of deliberately constituting nature, science withdraws humans from it, in effect reintroducing the distinction between the "real and objective" and the "merely subjective." Even more curious is how "real" becomes defined: not in terms of actual individual and social experiences of physical, biological, psychological, or social nature but of theory-dependent data.

The idea that nature is both anthropologically and epistemologically constituted is especially revealed, however, in reflection on those parts of nature that we have more fully and consciously appropriated. Every time we look at a field of wheat, we delude ourselves into thinking that we are looking at nature. But if we look a little more carefully, we are forced to realize that the field of wheat is not the product of nature but of culture. In the most obvious sense, the very fact that there is a field of wheat (with little else in it), and not something else, bears witness to the cultural character of the wheat field. Indeed, that is why we speak of agri*culture*. Moreover, at some time, some group of people identified what we now call wheat as food; over thousands of years farmers have shaped and reshaped the wheat plant so that it would be shorter, have stiffer straw, be more pest resistant, yield more, and make better flour and bread.

Bannerot (1986) argues that proto-varieties of food crops began to appear as soon as gatherers began to plant; they avoided thorny or bitter varieties and replanted those that exhibited a hypertrophy of the edible parts. Proto-varieties might even have emerged earlier, before gatherers learned to plant. Similarly, Chapman, Delcourt, and Delcourt (1989) have shown how Native Americans transformed the landscape of eastern Tennessee long before the coming of Europeans; indeed, the apparently primeval view that naturalist William Bartram saw in 1776 was actually the result of centuries of human-induced change.

In the last century plant and animal breeders have taken what farmers developed and used experimental methods and genetic theory to change plants and (to a lesser extent) animals even more radically. Indeed, plant and animal breeders, it may now be argued, have virtually monopolized the process of socializing living nature in the last century (Chatelin and Riou 1986). The same process has probably changed the trees that line our streets, the flowers in our gardens, and even the insects that eat our crops and the diseases that befall them.

Reboul (1977:87) has gone further. He argues that even soils are the product of society rather than nature: "To the extent that soils are at the same time the causes and consequences of systems of culture and livestock raising practices, their fertility depends directly on social categories that assure their improvement, as well as on the social and economic system that founds them."

When we speak of domesticated plants (or animals), we refer to a nature that we have created, a nature that is a material manifestation of descriptions, evaluations, and manipulations, all reflective of how a society constitutes nature.

Just as we create social structures by participating in them, that is, by creating networks that link together otherwise heterogeneous things, including inanimate objects, living things, and people (Bijker, Hughes, and Pinch 1987), so we make nature by participating in it. By socializing plants and animals, we make them conform to an order envisioned and valued by human beings. Indeed, this is perhaps the key feature of *homo faber*: by making nature, particularly in the production of food and fiber, we produce and reproduce ourselves as well.

Of course, it could be argued that wheat plants are somehow different from other humanly created products such as tractors because they are living things. But this in no way makes them less our creation than are tractors, nor less a part of the system of manufactured nature. If a tractor is not maintained, it will soon become useless. Moreover, one's ability to maintain it is inextricably linked to a network of tractor manufacturers, supply firms, gasoline and motor oil distributors, and repair shops. Without them, the tractor is merely a shiny ornament of no value for its intended use.

The same is true of the maintenance of cultivated plants and farm animals. If plants are not nurtured in the field, if the land is not plowed, if plants are not fertilized and irrigated, weeded and harvested, they soon turn into a pile of lifeless forms. Similarly, if dairy cows were not fed, milked, cleaned, and cared for, they would soon die from disease and hunger. As there is a network of people and things to support the use of tractors, so there is a network to support the use of particular varieties of plants and animals. In Western societies this network consists of seed producers; fertilizer, pesticide, feed, and antibiotic suppliers; irrigation equipment dealers; milk collection systems; railroad and truck transport; purchasing agents; food processors; wholesalers; and retailers, among others. In peasant societies it consists of exchanges with neighbors, local markets, local recipes, and other elements of local culture.

Cultivated plants and farm animals are as much cultural artifacts as are machines. Their cultivation is dependent upon and intertwined with a host of other cultural artifacts, including institutions. Domesticated plants and animals are only the most visible material aspects of social decisions and social structure. Hence, plants and animals fade into nonuse, even oblivion (as is the case for many landraces that are no longer cultivated), as soon as we cease to care for them, as soon as the social structures of which they form a part change.

Therefore, the maintenance of a diversity of crop plants is intimately linked to the maintenance of diverse societies. Put another way, the maintenance of cultural diversity in the field is part and parcel of the maintenance of *human* cultural diversity. Because of the inextricable link between agriculture and social structure, the choice of what domesticated plants to continue to cultivate and what plants to consider cultivating is a matter of great social concern. The problem of the conservation of plant germplasm[1] for actual or potential cultivation is as much a problem of what form social structure shall take as it is one of what technical products and procedures should be used.

Some Thoughts on the Sciences

Science, like all human endeavors, is a social institution. Scientists belong to communities that include not only other scientists but interessees (Radnitzky 1973) or clients (Busch and Lacy 1983). Scientists, at any given time, are able to work on several projects, to pursue several research trajectories, to investigate many aspects of nature. For example, wheat breeders might breed for yield, disease resistance, insect resistance, drought tolerance, grain quality, or earliness, but they can never investigate all possible paths. Thus they must choose between multiple goals. For example, they can rely on the work of other scientists in defining priorities, do the easy task, do what they find interesting, undertake those projects for which funds are available, or address (potential) clients' needs. Probably all of these and other considerations come into play in most decisions as to which problem to pursue.

For example, proponents of the Green Revolution decided, independently of the concerns of poor farmers, that higher-yielding production systems could be developed for the Third World. Only after the new technologies were in hand were farmers asked for their thoughts on the subject. Many were more than happy to adopt the new technologies, while others rejected them as useless. What was supplied was useful for some but useless to others, desired by some, rejected by others.

Turning to the demand side of the equation, it is apparent that (potential) clients have various things or processes that they desire or need. These are demands for hypothetical products, for things that do not yet (or may never) exist. These hypotheticals form the "demand" for science and other social changes. Krohn and Schafer (1983) have noted that Justus Liebig developed the field of agricultural chemistry in response to societal demands. Agricultural chemistry was to be a chemistry applied to the problems of agriculture, specifically to the problems of plant nutrition. Of course, not all farmers but only a select few urged the development of agricultural chemistry. The same would doubtless be true of other demands for science because the needs of clients are not (normally) homogeneous. They vary greatly by status, class, gender, ethnicity, region, and in other ways. Moreover, some clients are more articulate than others in expressing their needs or desires. Finally, only a small portion of the needs of clients can be addressed through science and technology.

Ordinarily, only a few of the many demands made by various publics are the subject of scientific investigation. There are numerous reasons for this. For example, a given client group may be unable to transform its demands into terms scientists can understand. Subsistence farmers in developing nations, for example, are often unable to articulate their demands to scientists. Alternatively, the problem may be well understood but there might be general agreement among scientists that it is (presently) insoluble. Perhaps the problem simply is not amenable to scientific analysis because of its complexity or even the way the sciences are organized. Until recently, this was thought to be the case for the analysis of complex agricultural systems. Or perhaps scientists consider the problem to be someone else's and not central to their discipline(s). Thus during the controversy over DDT, many entomologists saw the death of wildlife as a problem outside their domain (Dunlap 1981). The problem could be of low priority to the scientific community. Perhaps the clients who demanded the research are not the ones who will use it, and those who might use it see little of value in it. The resolution of the problem might be politically discomforting because a more powerful client group prefers to maintain the status quo. Or perhaps funds are unavailable (or are available in insufficient quantity) to accomplish the task at hand. For these and other reasons, only a small portion of the demands made by clients will be translated into research.

Similarly, many of the things that scientists are able to supply never extend beyond the scientific community. For example, scientists may be furnishing solutions to problems that are of little or no concern outside the scientific community. Alternatively, they may be able to supply a new technology but at a price that is far above the willingness or ability of clients to pay. Perhaps the products that scientists supply are hard to use, difficult to implement, or even dangerous. The use of the product could also be considered unethical, immoral, or sinful. Perhaps what works just fine in the laboratory cannot be scaled up to industrial production levels. Or maybe the use of the product would prohibit users from engaging in other, more important activities. It might even be rejected because its apparent effects on social relations would be too upsetting to the status quo.

Many of the things that scientists can supply are of no interest to clients. Similarly, many of the things that clients demand are not likely to be supplied by scientists. For ordinary commodities that have material form and are homogeneous supply and demand converge within a market. Because the products and processes of technoscience are hypothetical, it is only through complex processes of negotiation, persuasion, and coercion that supply and demand converge.

One would hope that the history of science would show numerous examples of these negotiation processes. Yet it is still dominated by the Great Man theory. Thus historians often tend to write as if single individuals were able to develop whole fields of science by sheer perseverance. They are also able to argue – with the advantages of 20/20 hindsight – that the now established facts could not possibly have been otherwise. What is concealed by the Great Man theory is that not only the facts but the very nature of the problem are defined through negotiations. Consider, for example, the current debate over global warming. On the one hand, numerous scientists have accepted global warming as established fact and are beginning to examine the likely consequences. On the other hand, numerous scientists find the evidence for global warming unconvincing. In short, the existence of global warming as a phenomenon worthy of study is still in dispute. Such definitional negotiations are almost never discussed by historians who subscribe to the dubious Great Man theory.

In addition, students of the history of science tend to be faced with stories of all the successes of scientific endeavor and few of the failures. Lost is the enormous portion of scientific research that is socially defined neither as a success nor as a failure, including scientific literature that is never cited at all, never referred to by other scientists, or never used in the development of new technologies. In addition, there are many new technologies and products that are never used or are rejected by users soon after they try them.

Nevertheless, some outright failures are both necessary and useful. Trial and error still constitutes much of what occurs in any scientific laboratory or field. Scientists are tinkerers who jerry-rig apparatus in the lab, trying various ways to get things to work (Ravetz 1971; Clarke and Fujimura 1993). Machiavelli (1985) was correct in arguing that chance or fortune plays an important role in the affairs of human beings. Social scientists and even many politicians today ignore this important aspect of human existence. For example, Thomas Edison tried hundreds of materials as candidates for filaments for his electric lights. Only by sheer perseverance did he succeed in finding one that worked. Similarly, scientists at large chemical company laboratories today are using tissue culture to test hundreds of varieties of maize so as to identify those few that are resistant to glyphosate. Those, in turn, will be used to develop herbicide-tolerant crops. In experiment stations around the world, scientists are testing new cultivars (i.e., cultivated varieties) in field trials. Such trial and error is essential to scientific and social progress.

Starting Points

To summarize, nature is made and remade by human beings through a wide variety of processes that may be described collectively as culture. Thus plants, animals, and machines are all manufactured (i.e., made by human hands) through culture.

Technoscience[2] is a major part of our contemporary culture and plays a major role in reshaping both nature and culture. Yet the technoscientific community often is unaware of how its preoccupations with nature are both guided by and, in turn, transform the cultural world. Thus however much it may seek to be value-free, technoscience brings values and interests into the process of making nature.

Technoscience must take into account the necessary links among three aspects of culture and nature: food, agriculture, and environment. In the recent past we have kept food, agriculture, and environment in separate categories, both conceptually and in practice ignoring the myriad ways in which they are related. Hence those concerned with agricultural production have paid relatively little attention to food and nutrition issues. Those who have been concerned with the environment have virtually ignored the role played by agricultural activities in shaping and reshaping the natural environment. Similarly, those concerned with food and nutrition have cared little about food production per se.

This separation has led us to a certain blindness with respect to the innumerable interrelations among the three domains. Yet if nature is not natural, if we make and remake nature by our actions, then agriculture is surely one of the most important and long-standing ways of accomplishing that task. Today, farmers and environmentalists are often at odds, yet they have every reason to be allies. The good agriculturalist cares for the land, nurtures it, conserves it. The good agriculturalist is concerned not merely about the short-term profits to be made from the sale of crops – though these are doubtless of importance in a capitalist society – but also about the communities that are sustained by agriculture.

Similarly, those concerned about the environment cannot limit themselves to the conservation of the wilderness that covers less than 5 percent of the total surface area of the planet (Pimentel et al. 1992). The "forest primeval" is not the only source of pleasure one experiences when passing through the countryside. Cultivated fields and towns and villages – what Hiss (1989a, 1989b) has called working landscapes – are also valued, even though the human element is much more visible in them than in the forest. Attempts to restore the landscape to some earlier form, such as the restored prairies in the Midwest, must necessarily retain the human touch (Turner 1988).

Similarly, those concerned about food are concerned not merely with its abundance and nutritional value but with its origins and meaning. Consequently, food produced in the backyard garden is often viewed as better than that bought at the store; prestigious foods are eaten at weddings and other festive occasions; and many entirely edible plants and animals are not eaten even though they may be found in abundance.

In short, food, agriculture, and the environment need to be reconstituted as a whole again, as part of a complex system of which humans, plants, and animals are a part. Within this system we coevolve and our technologies change us as much as they change the rest of nature.

Most technoscientific decisions are also ethical decisions. They involve the (re)making of nature to suit human purposes and interests, and they have consequences that affect people and their environments. The ethics of scientific work is important because such work is multidimensional. On one level, scientists simply find ways to transform nature for specific ends. They wring an extra bushel of grain from an acre. They separate, mix, refine, restructure, move, mold, heat, and cool nature. Yet on another level, each of these apparently mundane transformations of nature also implies a transformation of us as human beings and of human relations. This transformation is manifested in two ways. First, the ways we change nature reveal the kind of people we are. If we deliberately transform the beauty of nature into a wasteland, it is not just that something beautiful has been destroyed. Something within us has also been destroyed. As surely as living in a hovel reflects one's poverty and living in a mansion one's wealth, so the relations we have with nature reflect our commitment to other values. Second, the ways we transform nature change our social relations. The transformation of nature, in technoscience and in other work, is a collective enterprise. Thus new ways to remake nature must necessarily involve new forms of social relations. A wheat cultivar that matures earlier requires that people be available to harvest it at a new time. Larger farm equipment requires fewer workers in the field. Assembly-line technology such as that used on the tomato harvester creates new forms of social control. Automation makes whole occupations superfluous. There is no way to avoid this. Every technical change brings with it a corresponding social change; the two are separable only conceptually.

Agronomist Boysie Day is one of the few scientists to have noted the inseparability of technical change and ethical values. He argues that the migration to the cities, the displacement of labor by the cotton harvester, and other social changes are the direct result of the technical changes created through the quiet revolutions started by agronomists. He goes on to argue that "we fear that when his [i.e., the agronomist's] values come in conflict with questions of political expediency, social justice, and equity, the production ethic takes precedence. Such is the stuff of revolutions" (Day 1978:26). What Day fails to recognize is that the values of technoscience need not be blindly wedded to what he refers to as the "production ethic." Indeed, moral responsibility to the larger society demands that scientists take a broader view of their work.

Biologist Martha Crouch has noticed the same connections. She asserted that technoscience often embraces those who are powerful and pushes aside those who are powerless. Many of the current environmental problems that plague us have their roots in technoscience. She concluded that "using science and technology to solve problems in the world has gotten us into our current situation and is, thus, not the approach for getting us out" (Crouch 1990:275). Yet abandoning technoscience would hardly get us out of our present dilemmas. We have taken on responsibility for the world by our numbers and our technologies; we cannot return to some pristine past in which agriculture was everyone's work and technologies were less intrusive. There is little choice but to face up to those responsibilities that we have placed upon our own shoulders (Jonas 1984; Burkhardt 1990).

Technoscientists in the public sector, whose obligation is to serve the public good, cannot create new products and processes without considering public good questions, for every technical change has consequences that may or may not serve the public good. In practice, most technoscientists already consider public good questions to some extent. Technoscience is expensive, and the agencies that fund it demand that certain goals be met. Yet it is rare that these goals are perceived in terms of the larger questions of social ethics or the public good: Will the new products or processes serve to decrease inequalities? to increase freedom and liberty? to enable the less fortunate? to provide a sound environment? Or will they serve to create new class barriers? to put new restrictions on people? to degrade the environment?

Admittedly, the answers to these questions are not simple. The outcomes are not always apparent at the inception of a project. Yet scientists always have far more potential avenues of research to follow than time permits. The challenge is to choose those that are most likely to promote the public good and to monitor those choices so as to improve the selection process in the future (Friedland and Kappel 1979).

The Plan of This Book

In the following chapter we examine the growing homogeneity of certain aspects of human culture and show how it relates to the growing homogeneity among cultivated plants. In particular, we trace certain aspects of the history of both food and agriculture, showing how the two were separated, how the world was won (for the West) by economic botany, and how the world was made one by that same science. Chapter 3 focuses on institutions and debates that characterize the current international debate over plant germplasm. The next three chapters examine the situation in the United States. Chapter 4 examines the history of germplasm conservation and related activities in the context of a changing U.S. agriculture. Chapter 5 provides an overview of the U.S. National Plant Germplasm System, and chapter 6 looks at the nongovernmental organizations – both nonprofit and for profit – and asks how they contribute to this task. The cases of France, Brazil, and Chile provide important contrasts to the United States. These three nations have different cultures and different agricultures from those in the United States. They are the subjects of chapters 7, 8, and 9, respectively. The international debate over property rights in plant germplasm is the focus of chapter 10. As we shall see, that debate both establishes and limits the options for achieving solutions to the difficult problems facing us all. Finally, in the concluding chapter we ask what kind of nature we want and return to the problems associated with constituting nature. Our hope throughout is that as we continue to go about making nature, we begin to do so with explicit concern for how we affect people and their environments, immediately and far into the future. Our ultimate hope is that in the matter of germplasm, as in all matters, we begin to reestablish a culture of care.

2

Properties of Culture

> None of us have crystal balls to look into the future. What we need today, and think we need today, may not be what we think we need tomorrow.
>
> Technician at a U.S. university

> There is an ironic paradox regarding breeding programs – vegetal and animal alike: the same tools that provide the improvement are the tools that make genetic vulnerability possible.
>
> Scientist at a Brazilian research agency

It is a truism that in today's multicultural world peoples who are biologically of the same species may live out their lives in radically different ways. Yet events of the last few centuries – a minute portion of our brief human history – threaten to transform the world in ways that are both hazardous and unexpected. Until relatively recently in human history, the vast diversity of human cultures was largely unknown. Members of each culture lived within relatively small spatial contexts and had only occasional contacts with neighbors. Doubtless, each of these small human groups was convinced that it represented the right, the proper, the correct, and the only reasonable behavior, while those who were members of other groups were not quite fully human or were barbarians at best. Thus the diversity of human culture was arguably at its peak when most human beings were unaware of and unconcerned about it.

Starting with the rise of empires in China, Peru, Egypt, India, and elsewhere, however, the awareness of foreign cultures increased. As the empires increased in size and power, local cultures were forcibly eliminated and replaced by the encroaching civilizations. All this, too, is well documented and needs no lengthy discussion here.

In recent times, beginning with the great explorations of the fifteenth century, contacts between cultures increased dramatically. The empires of the distant past were eclipsed by the world empires of the eighteenth and nineteenth centuries, and the capitalist world system (Wallerstein 1974) began to cover the globe. The system of nation-states, brought into existence by the Treaty of Westphalia in 1648, has gradually suppressed and eliminated many, if not most, of the local cultures of the world. Traditional nomadic peoples were the first to go, often literally swept from the face of the earth by states that could not count, tax, control, instruct, or rule those who move. Whether we talk of Native Americans in the U.S. plains, nomadic herders in the Sahel, Gypsies in Europe, Amazonian peoples, or the desert societies of Mongolia, the story is essentially the same.

We have now arrived at a point when virtually no culture is still unaffected or untouched by contact with the larger world. Moreover, the older cultures are rapidly collapsing, either through direct pressure from powerful states or from within, from the increasing irrelevance of the collective knowledge that they maintain.

In the last few decades, we have reached yet another stage in world history in which a homogenizing process is occurring on a global scale. As a result first of improvements in transportation and then of instantaneous mass communications, the world has not only become smaller; it has begun to lose much of its cultural diversity. Nomads cross the Sahara with small radios tuned to the latest world news. The residents of remote Indian villages gather around a color television set and watch a videotape of a recent cricket match in England. Cities as far apart as Singapore, Abidjan, and Toronto resemble each other. Glass and steel skyscrapers, luxury hotels, McDonald's restaurants, and Coca-Cola span the globe.

Several options are now open to the world. First, we might let the homogenization process continue unabated. We might produce a world in which there are no surprises, in which security and stability take precedence over local cultures, in which local knowledge ceases to have any reason to exist.[1] Second, we might let the world degenerate into a kind of high-technology tribalism in which nationalist and nativist movements continually erupt violently, at considerable cost in human lives. Indeed, this might happen in spite of homogenization. Alternatively, we might create a pluricultural world in which local knowledge is maintained, created, and modified alongside knowledge that is global in scope. There are perhaps other alternatives as well.

But what does all this have to do with germplasm conservation? The answer is surprisingly straightforward. As we shall demonstrate, the concern for germplasm conservation is a material manifestation of the homogenization currently under way. To date, those proposing solutions have accepted homogenization as inevitable if not desirable. Yet the homogenization of culture implies that all of us will demand the same products and services, the same food prepared in the same ways, grown in fields cultivated in the same ways. Thus many of their solutions have served to exacerbate rather than to ameliorate the problem. Perhaps the only way to conserve cultural diversity in the field is to conserve cultural diversity among peoples. After all, if, as we have argued above, the nature we know is ours, then its homogeneity or heterogeneity is merely a materialization of culture.

In this chapter we ask two interrelated questions. First, we ask how the world was won. How did we arrive at the point at which control over nature was centralized in a few places? Then we ask how the world was made one. How and why did the problem of declining genetic diversity arise? We ask both of these questions with respect to agriculture, food, and environment, for as we shall see, the three are inextricably linked throughout history. Finally, we look briefly at the complex problems associated with germplasm conservation today.

How the World Was Won

Botanical gardens have been a part of human culture since at least 2000 B.C. (Juma 1989). But it was not until the late fifteenth century that modern botanical gardens arose as part of what has been called either the Age of Discovery or the Age of Imperialism. Although the colonization of the world by the Western powers is well known, the transfer of seeds, plants, and other genetic material is often forgotten (Juma 1989). Yet, Christopher Columbus brought seeds with him on his voyages to America. In 1524 Charles V was informed by Hernán Cortés of the importance of seeds in the colonization of the New World (Plucknett et al. 1987). In 1526, Gonzalo Fernández de Oviedo produced the *Summary of the Natural and General History of the Indies,* describing the plants of the islands, including maize (Casa Valdés 1987). And in 1570, Philip II sent the court physician, Francisco Hernández, to the Americas to bring back medicinal plants. During the seven years he was there, Hernández collected, cataloged, and brought back to Spain 400 samples (Casa Valdés 1987).

Indeed, now, 500 years after Columbus's voyage across the Atlantic, it is hard to remember that one very compelling reason for the trip was to obtain spices and silks from the Orient, that is, to obtain, test, and transplant rare and unusual agricultural products. The voyages of the European explorers extended the boundaries of empire around the world. They radically altered the diet of nearly everyone on the planet. Moreover, they began to restructure the global environment.

The scope of the gardens is one example. The first modern botanic garden was founded by the Venetian Republic at Padua in 1545. The Venetians were particularly interested in learning more about plants so as to further their trading position (Prest 1981). The British gardens at Kew, the Jardin des Plantes in Paris, the Amsterdam Botanical Garden, the Jardim Botanico in Lisbon, the Botanical Garden of Aranjuez (founded by Philip II but no longer in existence), and the Royal Botanical Garden in Madrid, among others, formed the basis for the transfer of plants from Europe to the "neo-Europes" (Crosby 1986) in the Americas and Australasia (e.g., wheat, oats, carrots), from the temperate zones of the world to Europe (e.g., potatoes, maize, tomatoes), and from tropical and subtropical colonies in one part of the world to similar agroclimatic zones in other parts of the world via Europe (e.g., rubber, tea, coffee, cocoa, tobacco).

At the same time, gardens were established in each of the colonies. Brockway (1979) has estimated that as many as 1600 gardens existed by 1800, connected into a dense network making possible the systematic knowledge of the world's plants, their uses, and cropping systems in the capitals of the European powers. Eighteenth-century economists, and the physiocrats in particular, developed a theory that saw agriculture as essential to creating the wealth of nations. "Every new plant was being scrutinized for its use as food, fiber, timber, dye, or medicine. Botanic gardens consciously served the State as well as science, and shared the mercantilist and naturalist spirit of the times" (Brockway 1979:74–75).

The colonial botanical gardens provided a worldwide system for the exchange and evaluation of plants. Those with economic potential were sent to similar climatic zones around the world for testing. For example, Kew Gardens in Britain served as a plant evaluation and transfer station for thousands of plants collected throughout the British Empire. Sixty-three years before Sir Henry Morton Stanley's famous trans-African expedition, the British had sent a scientific expedition up the Congo River with a gardener from Kew (Cornet 1965). They also sent the now infamous Captain William Bligh on the HMS *Bounty* to bring breadfruit trees from Tahiti to the West Indies in 1793 to provide food for slaves working on sugar plantations there. Bligh's crew mutinied in part because the limited fresh water on board was being used to water the 1000 trees collected. The British also took rubber trees from Brazil and used them to establish huge rubber plantations in what was then Malaya. They also introduced potatoes to New Zealand and Ireland. In the latter case, the varieties of the new crop that were introduced had the great advantage of permitting much higher yields of calories per unit of land than the cereals previously cultivated, thereby making possible the enclosure of larger land areas for animal agriculture. Combined with the firm belief in laissez-faire economics that prohibited relief, these events contributed to the tragic Irish potato famine of the 1840s.

The French Jardin des Plantes (see chapter 7) served as a model for similar gardens in Réunion, Martinique, and Mauritius (Haudricourt and Hédrin 1987). The French moved grapevines and commercial citrus groves from the gardens to Algeria (Brockway 1979). The Portuguese used the Jardim Botanico in Rio to test dozens of foreign economic plants – so many that visitors complained of the lack of native flora. Among the many plants imported was tea (along with 200 Chinese laborers) to establish plantations in Brazil. And Spain's Jardim de Aclimatación was established in 1758 in the Canary Islands in an attempt to adapt tropical plants to the European climate (Smith 1985).

These gardens made possible the transfer of desirable cultivars over long distances. New technologies such as the Wardian case (Juma 1989) – a kind of portable greenhouse – were developed to help plants withstand the long ocean voyages to move them to new ecozones. Diets became more diverse and were vastly improved as crops from the Americas were imported to Europe and its colonies in Asia, Latin America, and Africa, and crops of those regions made their way to the Americas. By assembling plants from around the world in one location, the gardens permitted the refinement of a global botanical classification system that allowed identification of plants of like genera. As Latour (1987) suggests, the gardens made it possible to take a trip around the world in one room. In addition, they permitted the development of large plantations upon which great monocultures could be grown, replacing the diversity of local agroecosystems. All this probably eroded the germplasm base upon which we depend, but the effects were initially small enough to escape notice.

The great plant transfer also changed food in innumerable ways. The ingestion of food occupies a peculiar status in human societies because it is foreign matter entering the body. If nature was the unknown, then food represented a fundamental way to communicate with it. Food consumption in all societies is not merely a matter of ingesting nutrients but of symbolism and meaning. Christian imagery of bread and wine, the prohibition against pork in Judaism and Islam, and that against beef in Hinduism are all part of the complex and varied ways that food permits a communion with nature. Moreover, before the eighteenth century for nearly everyone, food production and consumption were intimately linked; what happened in the fields was inextricably tied to what happened in the kitchen.

Even when it is possible to do so, food is almost never consumed without some sort of transformation. The raw is turned into the cooked not only because cooking aids digestion or makes the food easier to consume but because cooking removes pollution and enhances purity (Lévi-Strauss 1969). The transformation of food has proceeded alongside that of agriculture, but it has taken several paths. First, agriculture and food have been institutionally separated. Second, certain aspects of food selection and preparation have been removed from the kitchen. Third, the kitchen itself has been transformed so as to make it conform more satisfactorily to the new norms.

THE SEPARATION OF FOOD PRODUCTION AND CONSUMPTION

Anyone who has ever grown, cooked, and eaten vegetables knows the pleasure of eating that which one has produced. Until the eighteenth century virtually everyone had this small pleasure. Farmers might be forced to work for others, but their knowledge was entirely their own and could not be easily appropriated by the ruling class. Similarly, people might cook for others, but the ruling class would not appropriate the knowledge of cooking itself. This situation endured until the great transfers of plants began.

Then, with great rapidity – given the transport and communications of the day – cropping patterns, domestic animal populations, and diets changed. The combined effects of the enclosures, industrialization, and specialization among farmers began the still ongoing severance of farming from food preparation and the decline of local knowledge of both. The African case is a good example. Until the Columbian exchange (the worldwide transfer of plants and animals begun by Columbus), Africans living south of the Sahara subsisted on a diet of sorghum, millet, and yams. The exchange brought maize, sweet potatoes, peanuts, and cassava. Even before the colonial period, maize became the most widely cultivated food crop in Africa (Davis 1986). Maize yielded much more than either sorghum or pearl millet, and cassava could be harvested throughout the year and was more resistant to locusts and drought than other locally available tubers. The land and labor freed as a result of the higher yields were often used to cultivate peanuts and oil palms for export to the north.

Colonization increased the pressure on the population to produce cash crops such as coffee, cocoa, rubber, and cotton. Often peasants were required to pay taxes in cash, which ensured that everyone engaged in export production at least to some extent.

Later, three other approaches were attempted with varying degrees of success. In some areas, large plantations were established. The plantations displaced farmers and often took the best land available to establish huge estates requiring thousands of workers. Of course, for those workers, food was now separated entirely from agriculture; they would henceforth buy their food in stores rather than grow it themselves. Moreover, they would come to prefer wheat bread, the food of the European colonists, over locally produced cereals.

In Europe, the enclosures had already forced people off the land and made possible the separation of food consumption and agricultural production. In Asia and Latin America, huge plantations were established to grow tea, coffee, rubber, tobacco, cocoa, oil palm, jasmine, and various spices.

In still other areas, complex systems of smallholder production for export were established. For example, in the Sudan, the world's largest single irrigation scheme – over one million acres – was established to produce long-staple cotton for the Lancashire mills. Here, too, agricultural science, the state, and the growing world trade in cotton were linked from the start. Indeed, one sympathetic observer noted "the factory-like appearance of Gezira farming" (Gaitskell 1959:174).

Finally, in a few places where agroclimatic conditions approximated those of Europe, "neo-Europes" were established. Native plants, animals, and often peoples were displaced or eliminated in favor of European equivalents (Crosby 1986). Animals were often released to fend for themselves (either deliberately or accidentally), thereby creating a familiar source of protein for the European immigrants. Then European crops were planted, displacing traditional ones. In addition, because the indigenous populations of the neo-Europes were often small (or were reduced by European diseases and war), the neo-Europes were marked by high labor costs and consequent mechanization. In contrast, the areas devoted to plantation agriculture were usually marked by low levels of mechanization as a result of the cheap labor that was (made) available. In addition, often, when the size of the neo-Europe was small, the indigenous population was displaced and then prohibited from growing the European crops so as to ensure a market for the new settlers. This tactic was employed in the Kenyan highlands, for example.

For the nations that now make up the Third World, the result of the separation of food production and consumption was often catastrophic. Famines of a new sort developed in these nations. Unlike the famines of the past, when those who were starving or malnourished were relatively isolated from those with extra food, the new famines often occurred when people were forced onto more marginal land because land cropped only once a decade was now cropped annually and the alleviation of the consequences of urban unemployment grew beyond the capacity of village and family ties. In the industrialized nations the same problems continued to exist but on a much smaller scale. Moreover, they were partially compensated for by higher levels of social services won through class conflict. Nevertheless, the not so gradual break between food consumption and agricultural production continued unabated.

NEW ENVIRONMENTS

The changes in agriculture described above caused changes in the natural environment. As agriculture has become more specialized, it has restructured the rural landscape. Tree crops harvested in the wild are now planted in rows over immense expanses. Garden crops, such as tomatoes, are now planted in huge fields as monocultures.

The variation within each crop species also appears to have decreased. There are two interrelated reasons for this change. On the one hand, high-yielding varieties (HYVS) have replaced the traditional landraces of millions of farmers. On the other hand, the standardized raw materials needed by the processing industry have demanded greater uniformity in the fields. Moreover, the area under cultivation has increased. Unspoiled forests and other areas not previously tilled have come into use. In addition, the best land has often been taken for the production of industrial commodities while poor farmers have been pushed onto more and more marginal lands. The Dominican Republic offers a case in point. Much of the flat, fertile land of the valleys and seacoast is devoted to production of sugar and of fruits and vegetables for export. Smallholders have begun to colonize the hillsides and are gradually denuding them. In neighboring Haiti the process has gone so much further – though for different reasons – that the boundary between the two nations is visible from the air. But the transformation has also moved into the kitchen.

REMOVING DECISIONS FROM THE KITCHEN

In 1795 the French government offered the prize of Fr 12,000 for the invention of a practical method of preserving food for the army. In 1809 the prize was given to Nicolas Appert, a Parisian confectioner, who was able to preserve foods in glass bottles. Initially, this approach was confined to supplying

Napoleon's army, but it spread rapidly to civilian use, particularly for out-of-season products.

When canned and frozen foods were introduced, it became much less clear just what was being purchased. Contents labels replaced visual inspection, and although pictures were often printed on the labels, the actual contents frequently were of considerably lower quality than the pictures showed. Government agencies were established to ensure that what was in the package was accurately described on the outside, that contamination by foreign matter was reduced or eliminated, and that minimal health and safety precautions were followed. In the United States a 1938 law permitted the Food and Drug Administration (FDA) to specify contents labels for many foods. Ingredients were to be listed on packages in descending order by weight (for solids) or volume (for liquids).

Only a few preservatives whose origin was unknown to consumers were included in processed foods. Thus the ingredients of packages remained both comprehensible – fitted within the generally accepted categories of experience – and apprehensible – immediately recognizable as known substances. A few ingredients did begin to appear that were not apprehensible (e.g., BHT) and probably incomprehensible to most consumers.

The next phase in the transformation of food occurred when the grounds of discourse shifted away from ingredients to nutrients in the early 1970s. Under the new regime, nutrients were to be listed by quantity (in the United States always in grams although packages are in English measure) and percent of recommended daily allowances (RDAs) (Porter and Earl 1990). The introduction of nutrient labeling and the gradual downplaying of ingredients moved the discourse to include items that are clearly nonapprehensible (e.g., carbohydrates) and perhaps incomprehensible to most consumers. As a result, instead of choosing from a limited number of food groups, consumers were faced with the much more complex (and more mystifying) task of choosing foods based on nutrient content and contribution to RDAs.

The final stage in the transformation of food is now occurring as products are reconstituted in such a way that they are no longer comprehensible. The creation of fabricated foods treats agricultural products as raw materials or components to be used in the manufacture of food products. At the same time, the use of biotechnology to transform plants and animals is now making possible the creation of "functional attribute crops" that are particularly amenable to food fabrication. Such crops would be designed by incorporating genetic materials from other organisms (and perhaps eventually wholly new organisms) so as to maximize or optimize the production of wanted nutrients and chemicals.

In addition, the continued differentiation of food products has led to bewilderment in the supermarket. No consumer, no matter how well educated, can possibly afford the time to make rational choices among the 30,000 items on supermarket shelves. Thus the knowledge of appearances that guided food preparation for millennia is rapidly being eroded by the restructuring of the food industry. At the same time, consumers' local knowledge of qualities is being replaced with scientific knowledge of qualities in a self-fulfilling prophecy. As the food system becomes more and more complex, local knowledge of taste, texture, color, and flavor becomes less and less meaningful. No longer do the correlations between color and taste apply; the beautiful round red tomato may be all but inedible. No longer can freshness be judged by direct observation; foods in sealed containers must be judged fresh according to dating systems that are themselves the product of lengthy and continued negotiation. Similarly, no longer can one glance casually at foods and determine something about their nutritional value. Perhaps of greatest concern, the reconstituted food may be less nutritious, less adequate, than the traditional diet because it presumes complete knowledge of the nutritional needs of people and the nutrients found in food products. Thus food becomes merely a simulation fabricated by the food companies in the name of nutrition.

THE INDUSTRIALIZED KITCHEN

The same quest for efficiency and organization that transformed farming also transformed the kitchen. In particular, the appropriation and substitution that Goodman, Sorj, and Wilkinson (1987) note on the farm also occurred in the kitchen. Paradigmatic of these changes is the work of Fannie Farmer (1896). Until hers and other similar cookbooks were published in the late nineteenth century, cooking proceeded by guesswork and practical experience. A pinch of salt or a dash of pepper, a measure of flour or a spoonful of sugar was the common way in which ingredients were combined. The introduction of standard weights and measures transformed cooking from an art into a science. This simplified greatly the task of learning to cook, but it also moved the source of knowledge about cooking from older generations to the cookbook writers. Moreover, it required not merely that one follow the recipes but that one reorganize the kitchen around the new recipes. New measuring devices would have to be purchased, new rules learned, new procedures employed.

Concomitant with the changes in recipes was the move to create efficiency in the kitchen by redesigning it, by reducing the number of steps one had to walk, by standardizing the heights of countertops, the construction of stoves, and the design of tables (Giedion 1975). The open hearth would be replaced by the wood, coal, and later electric or gas stove.

But the kitchen was transformed not only in the household. Of perhaps greater consequence was the industrialization of food preparation through a series of steps including the development of large restaurants and cafeterias and culminating in fast-food restaurants. Numerous arguments have been given for the rise of the fast-food restaurant. A larger portion of the workforce is salaried. The distance people travel to work has increased. More women are working outside the home. Family size has declined. "These various phenomena overthrow food habits that have hardly changed over the centuries, their principal effects being the destructuring of meals and the reduction of time reserved for culinary tasks" (Saint-Albin 1986:19).

Whatever the relative importance of each of these factors, the words of a French observer ring true: "Eating a meal appears more and more as an impulsive and solitary act for which prepared foods permit rapid satisfaction" (Dugowson 1986:16). Fast-food restaurants not only speed the removal of food preparation from the home; they also create new standards that are global in character.

First, fast-food restaurants are able to ensure the same quality of food on a worldwide basis by integrating backward into production and processing. For example, McDonald's requires huge quantities of fried potatoes, but not just any potato will do. Its engineers have discovered that the maximum efficiency is achieved when Russet Burbank potatoes are used. Thus all McDonald's suppliers are required to furnish that particular potato variety (Doyle 1985b). Similar requirements are attached to tomatoes, lettuce, hamburger buns, and the meat itself. This would be of little consequence if McDonald's were not the world's largest single purchaser of potatoes. In short, the uniformity of the food in the restaurant can be achieved only by creating equal uniformity in the field.

In addition, fast-food restaurants are spreading rapidly around the world, creating a homogenized global culture. In developing nations, fast-food restaurants are not where one goes to get a cheap meal. They are usually the preferred choice of the elites, the validation of one's modernity, upward mobility, or entry into the world culture. They are another indicator of development.

In short, the very same processes that have restructured food and agriculture have also restructured the rural landscape around the world. The "amber waves of grain" described in the American song are a part of the new nature that we have created in the last 500 years. The world was won by this restructuring of agriculture, food, and the environment. But this was not all that happened. It was also necessary to make the world one. That was done through the improvement of the techniques of plant improvement.

How the World Was (Made) One

We may divide the history of plant improvement – what counts as such is itself the subject of debate – into five more or less distinct periods, each building on the previous one. The first period was marked by the creation of agriculture. Probably beginning with the clearing of brush from in front of dwellings, people learned that individual seeds could be planted to yield crops (Rindos 1980). Over the centuries farmers selected seed from the plants with the largest yields of edible parts for replanting the following year.

This type of selection is still practiced in much of the Third World. It has little effect on overall diversity because many varieties are identified for different microclimates, seeds are exchanged within a small geographic area, and selection techniques are based on characters that are immediately obvious such as color, size, and shape.

The second phase in plant improvement began in the eighteenth century, when commercial plant breeders began to appear. These persons were the first to separate the occupation of breeding from that of farming. The techniques they employed were much like those of farmers. Trial and error was widely employed, supplemented by careful searches for exotic materials, a task for which the average farmer had little time. By the middle of the nineteenth century commercial breeding as a separate enterprise was so widespread that Charles Darwin was able to base his theory of natural selection on the outcome of the domestic selection by breeders (Mulkay 1979).

The third phase of plant improvement began with the so-called rediscovery of Gregor Mendel[2] at the turn of the century. Unlike the centuries of breeding that came before it, the Mendelian approach offered the possibility of theoretically guided experiment. Moreover, unlike previous attempts at breeding, Mendelian genetics postulated the existence of "factors" (later called genes) that could account for the variation displayed to the senses. Put differently, Mendelian genetics posited the existence of primary qualities invisible to the naked eye that created what appeared. Mendelian genetics accelerated the rate of progress in breeding, but it also took the selection process away from farmers. A turn-of-the-century guide to wheat breeding encouraged farmers to undertake a breeding program (Carleton 1900). By the 1930s, such a project was entirely unrealistic (Fitzgerald 1990).

Using Mendelian principles, plant breeders discovered that they could improve plants rapidly. Significant increases could now be obtained in 10 to 15 years or less. Moreover, the rapid integration of nearly all farmers into the world economy encouraged the spread of these varieties at an unprecedented rate.

The fourth phase in plant improvement began with the development of hybrids. These cultivars represented yet another step toward the displacement of secondary with primary qualities. Moreover, although there is considerable dispute over whether hybrids display heterosis (hybrid vigor), these new seeds were of interest to a segment of the plant improvement community for other reasons (Berlan and Lewontin 1986a). In particular, unlike varieties, hybrids would not breed true so their seed could not be planted to obtain a crop in the following year. Seed became an input to be purchased off the farm on an annual basis (Kloppenburg 1988). In the words of Goodman, Sorj, and Wilkinson (1987), seed production (at least for hybrid crops) was fully appropriated from the farm. Henceforth, not only would farmers be excluded from the creation of new varieties; those who grew the hybrids would be excluded from their multiplication as well.

Huge areas of the midwestern United States were planted to a few varieties of hybrid corn in less than a decade. Forty years later, improved wheat and rice swept much of Asia in the so-called Green Revolution. Only then did a few voices begin to ask, What would happen to the old varieties when the vast majority of farmers adopted the improved varieties and when more and more land was cultivated as a result of population pressures or ill-conceived farm programs leaving little room for the preservation of that which was not cultivated?

Moreover, the development first of hybrids and then of patentlike protection for plants has changed the world of plant breeding considerably in recent years. What used to be a public sector activity has, since 1970, become a private sector function as a result of the passage of the Plant Variety Protection Act (PVPA) in the United States and similar laws in other developed nations. Essentially these laws permit for the first time the restriction of access to plant materials, especially seeds. Since their passage most large seed companies have been purchased by chemical or pharmaceutical companies. Today, these companies dominate the seed supply industry in most Western and a few non-Western countries.

In the United States many of these changes were justified as bringing efficiency, organization, and productivity to the farm (Hays 1959). The scientist-led Country Life Commission saw organization and efficiency as the clear path through which rural America would keep up with the rest of the nation (Country

Life Commission 1911). And when Taylorism became rampant in the factories of the nation, American scientists were busily attempting to increase the efficiency of the farm. That the processes that were then set into motion would transform the very values they wished to maintain went unnoticed by the reformers.

The last decade has been marked by what is likely the fifth and final phase of plant improvement: the advent of the new biotechnologies. The techniques of cellular and molecular biology have been incorporated into plant breeding, promising considerably more precise breeding strategies. Whereas Mendelian approaches rely on sexual reproduction and thereby include both desirable and undesirable plant characters in each cross, cellular and molecular approaches should permit the insertion (and expression) of individual desirable genes in already high-yielding varieties. These new technologies hold within them the potential to transform nature in ways far more profound than ever before, producing varieties much more genetically uniform than older ones (Busch et al. 1991). They have the potential to reduce the variation within a given field to a degree that some would argue is dangerous to our food security (Busch and Lacy 1984b). They even hold the promise of moving food production to the factory (Rogoff and Rawlins 1987). These new technologies mark the final step in the socialization of nature and its transformation into resources.

In sum, each stage in the history of plant improvement has been marked by changes in the ways in which we collectively make nature.[3] But at the same time, each stage has been marked by a growing awareness of nature as something we make. Our remote ancestors filled nature with spirits and gods. Nature was to be feared because it was populated by evil spirits; it existed outside the boundaries of the known. Yet the personifications of nature that made nature known as that which is unknowable were entirely the product of human imagination.

With the rise of modern science the anthropomorphisms used to describe nature were removed. At the same time, the goals of knowledge were gradually transformed from understanding to an explicit recognition of control (Leiss 1972). Nature was to be made more human by removing human imagery from it. It was to be restructured, reshaped, and recreated to meet the needs and desires – not to say whims – of human civilizations. Nature was to be viewed as mere resources, what Heidegger (1977) called "standing reserves," there and available for the taking by whoever had the power to take and transform them. Each stage in the history of plant improvement marks both the increasing ability to make nature in our own image and the increasing inability to find our image in nature. On the brink of obtaining almost Promethean control over the forms that we shall make nature take, at the moment when nature is more than at any moment ours, we find nothing recognizable in nature.

Indeed, nature has been replaced, almost imperceptibly, by the environment. The transformation in language is revealing because we define nature as the essential or constitutive character of the world and the environment as that which surrounds. We are a fundamental part of nature and it is a part of us. The environment, however, merely surrounds us. Like a cloak, it may be removed and even discarded.

Plant breeders have become the key actors in this socialization process. "The breeder occupies a doubly strategic place: the knowledge acquired by different disciplines, the information on output markets, production economies, etc. converges toward him. The breeder is nothing without the contribution of his colleagues, but his colleagues are (virtually) useless without him" (Raymond 1985:58). Producers must take into account all those in the chain after them and still find what they believe is the most desirable variety to grow. Transporters, processors, and distributors have an interest in whatever facilitates their tasks. Consumers may reject an innovation as too strange or embrace it as novel and unusual. Or they may not notice it at all. As two breeders have put it, "before everything, he [the breeder] must respond, in a very limited time, to all the contradictory demands above but, besides, he ought to have his own conception of the plant and project it into long-term developments" (Bannerot and Foury 1986:94).

To date breeders and molecular biologists have been reluctant to admit the peculiar nature of their work. We find that like other sciences, plant breeding is also Janus-faced (Latour 1987), but with a different twist. Breeding is first, while it is occurring, a design process. The idea is to create a plant that is bigger or smaller, more resistant to a certain insect's ravenous appetite, stands up to the strongest winds, tolerates severe drought, uses artificial fertilizers to produce bigger edible parts. To accomplish this, breeders create as much variation as possible. They note unusual plants in the field for further study and, perhaps, to incorporate into some future variety. Then, after much work in the field and the lab, the same breeders work hard to eliminate precisely those novel, unusual variations in plants that were so interesting at first (Dattée 1986). The design process is now over, almost forgotten. What counts is to stabilize the new variety so that it appears as if it could not possibly be otherwise, so that it appears "natural."

In short, the Janus face of breeding is that it is at first a design process and then a process of discovering what nature already provided for us. Hence the new variety becomes part of a network (if it is widely adopted by seed companies, farmers, and others) that appears as if it simply cannot be otherwise. The social choices that are necessary, the negotiation, persuasion, and coercion that are central to the breeding process when the new variety is being designed, are glossed over. The new variety is simply the best that can be found "in nature!"

When breeding is looked at from this perspective, the complex web of relationships between people and plants are seen as a part of social structure. This peculiar form of social structure, like all others, must be worked at continuously by breeders and by all those they can enroll (Latour 1987): seed producers, farmers, transporters, processors, retailers, input suppliers, and others directly connected with the transformation of a given plant into a consumable product. Moreover, it also involves those apparently less directly linked to breeding such as lawyers, bureaucrats, judges, toxicologists, advertising agents, chemical producers – even sociologists who study the diffusion of innovations.

Contemporary Issues

A central aspect of each approach to plant improvement to date is the availability of a diverse gene pool. Breeders – including farmers – require a diverse germplasm pool from which to select improved materials. Yet at the same time, these individuals consider themselves successful when a cultivar that they have developed is widely adopted. The result is to reduce the diversity available in the field. As Wilkes (1983:134) has put it: "The technological bind of improved varieties is that they eliminate the resource upon which they are based." This gives rise to two interrelated problems: genetic uniformity and genetic vulnerability. Trial-and-error techniques and even the organized transfers of the botanical gardens had little effect on either. Modern plant breeding techniques have begun to have this effect as they have become commonplace around the globe. The new molecular techniques are likely to spread even more rapidly and exacerbate the problem.

GENETIC UNIFORMITY

Genetic uniformity refers to the tendency for monocultural production to be not only of the same crop but of the same variety. Essentially, breeders use the current best cultivars or lines as parent stocks for the next generation of cultivars and then repeat the process. The net effect is a continuing reduction of the genetic base, which was initially none too broad (Kannenberg 1984; Macer 1975).

The use of Mendelian genetics gradually began to create uniformity on a scale previously unimaginable. "Scientific" agricultural management and technology development early in this century provided further impetus for reshaping agriculture and transforming nature. But it was not until the last several decades that the full import of this restructuring of nature began to pose serious problems. The vast improvements in science, communications, and transport that have recently occurred have made it possible to sow much of the world's farmland in high-yielding varieties, to increase the amount of marginal land under cultivation, to increase the intensity of cultivation, to develop and spread worldwide farm animals that can gain weight very rapidly, and to harvest timber rapidly from huge forests to provide furniture and housing in the West and fuelwood in the Third World. Moreover, the widespread adoption of high-yielding varieties of plants and animals restructures the world of microorganisms. Complex ecosystems are replaced by greatly simplified agroecosystems. Within these simplified systems, "it must now be accepted that the activities of plant breeders determine the distribution of virulence types of many major pathogens – particularly the air-borne, foliar pathogens such as the cereal rusts – and provide the 'substrates' which favour the selection and multiplication of pathogenic forms with new combinations of virulence" (Macer 1975:355). Hargrove, Cabanilla, and Coffman (1985), for example, found that many Asian rice breeders used a particular semidwarf female in their breeding programs. These plants were obtained through the International Rice Testing Program (IRTP) at the International Rice Research Institute (IRRI) in the Philippines. As they point out: "IRTP nurseries clearly have a tremendous potential to spread new genes. But IRTP also has the potential to spread the same genes – combined into hundreds of 'new' parents – across the rice-growing world. We are concerned about the cytoplasmic similarity of many popular farm varieties, of new varieties, and of female parents used in 1984 hybridization" (1985:9).

They go on to note that this uniformity increases the probability of plant pest or disease epidemics for which most Asian countries are ill-prepared. There is certainly substantial precedent for such an event: Victoria blight on oats, leaf and stem rust on wheat, southern corn leaf blight, bacterial speck on tomatoes, and alfalfa weevil and sorghum greenbug outbreaks were all traced to genetic uniformity (Kannenberg 1984:94). Harlan (1984) and others (e.g., Mooney 1983; Reichert 1982) have noted that similar uniformity is typical of most major American crop plants. Duvick (1977) has produced data showing the apparently narrow base on which most major U.S. crops rest. More recent varietal data are scarce because most governments have stopped collecting it (NRC 1993), perhaps because of pressure from large seed companies that want to keep market shares secret.

In addition, some have argued that the recent decline in the number of seed companies, occasioned by their purchase by agrochemical companies, has also contributed to genetic uniformity (e.g., Reichert 1982; Mooney 1983). They argue that companies have a greater incentive to breed a single average variety to sell in a large area than to breed many varieties each better suited to local conditions. Data to support or refute this argument are insufficient.

GENETIC VULNERABILITY

A second issue, genetic vulnerability, refers to the threat of extinction of landraces and wild relatives of crop plants. This threat is made very real by the increasing adoption of high-yield monocultures to meet growing world food demands, as well as by the loss of uncultivated wild relatives resulting from population growth and industrialization. Because breeders must have variation with which to breed, no matter what techniques they use, genetic vulnerability poses a serious threat to the breeding enterprise. This issue is particularly important in the United States because nearly all major crops grown here, including maize, wheat, soybeans, and cotton, are of foreign origin. Without access to "exotic" germplasm, the development of higher-yielding varieties will doubtless decline; the high yields of American agriculture would likely decline as a result.

There is widespread agreement over the major causes of uniformity and vulnerability. "The forces that promote uniformity, and, therefore, vulnerability, are market place forces and consumer preferences" (Wilkes 1983:138). In short, farmers operating in competitive markets wish to maximize their productivity. Seed is usually a very small cost relative to other inputs so farmers seek out the seed that will produce the highest yields. Because adoption of these varieties means abandoning traditional varieties and there is generally a "best" cultivar for a particular locale, adoption leads to both genetic uniformity and vulnerability.

Surprisingly, the scientific community has been slow to recognize these problems. Although Harlan and Martini (1936) first warned of its consequences in 1936, and Nikolai Vavilov established a genebank in Leningrad in the 1920s, little concern was manifested by most scientists. For example, Hymowitz (1984) has noted that many of the soybean varieties collected by P. H. Dorsett and W. J. Morse in Asia from 1929 to 1931 were lost because the U.S. Department of Agriculture (USDA) had no interest in their preservation. Similarly,

Wilkes (1984) has noted how a world collection of melon seeds was thrown away, necessitating recollection when a virus broke out several years later.

GERMPLASM CONSERVATION

Our remote ancestors began to maintain plant germplasm when they decided to keep certain seeds for the following season. Germplasm conservation and varietal development and improvement were thus inextricably linked at their inception. Moreover, because the conditions of cultivation and environment differed from one place to another, the unorganized community of interest of countless farmers had the combined effect of increasing the variation in plant material and creating what we now know as landraces.

Farmers in ancient Rome were advised to select the plants with the largest edible parts for use in the following season (e.g., Virgil 1982). Even as late as the early part of this century, the idea of selecting for yield per unit of land area was not fully understood (Berlan 1987). Moreover, for most crops, farmers still preferred to produce their own seed.

The development of hybrid corn changed the situation dramatically. Farmers began to use hybrid corn because of its resistance to lodging, higher yields, and a shortage of seed corn as a result of poor weather. But there was a price to pay: continued dependence upon the seed companies to provide seed for the following year's crop (Berlan 1987; Kloppenburg 1988).

When farmers stored all of their own seed on the farm (as they still do in many developing nations), many of the present-day problems of storage did not exist. Of course, farmers did their best to ensure that the seed was not attacked by insects, diseases, rodents, or birds. They tried to keep the seed dry and away from prolonged exposure to the sun. And they used the seed within one or two seasons so as to achieve high germination rates. These practices did not result in the survival of all seeds, but rarely was everything destroyed. Farmers who followed the time-honored procedures of their grandparents were usually able to survive another season and plant another crop.

Of course, many farmers did lose their entire seed supply through plague, pestilence, flood, or war. Local famines were a constant threat everywhere, and poor roads and communications made relief from famine difficult at best. In medieval Europe peasants often had a more precarious life than townspeople, for in times of short food supplies townspeople could search elsewhere for food, whereas peasants usually lacked the cash to trade for food. The situation was even worse in India and China. As Braudel (1973:38) put it, "Famine recurred so insistently for centuries on end that it became incorporated into man's biological régime and built into his daily life." Thus during times of famine farmers and their families often ate their seeds and begged seeds from neighbors the next season or did not survive.

The creation of world markets, global seed companies, and improved transportation and communication radically reduced the likelihood of local famine. But at the same time that it decreased local vulnerability to famine, it may have increased the vulnerability of the world as a whole. That is, if farmers in a small area died from lack of food, their germplasm died with them. This was a great tragedy for those involved. Millions of other producers with millions of landraces remained untouched, however. In contrast, today only a handful of institutions are responsible for maintaining the world's crop plant germplasm. If those organizations were to be rendered ineffective, people over a much wider area might suffer food shortages, famine, and death. Much of the redundance in the maintenance of the collections has disappeared.

Over the last several decades, in part as a result of pest outbreaks and in part as a result of improved communications, germplasm maintenance has begun to receive more attention throughout the world. Genebanks have been formed to store varieties no longer grown or on the verge of extinction.[4] The questions that each farmer faced in past times – which seeds to store? for how long? under what conditions? – have become institutional questions. Whereas an incompetent farmer's loss of his or her seeds would have only personal and familial consequences, incompetent, inadequately funded, or ill-conceived maintenance strategies today can have global consequences. Hence what was once a relatively technical question has been transformed into an ethical one.

The technical issues have not vanished, but are of far greater importance today than they once were, precisely because the ethical weight they bear has grown in magnitude. Making the wrong decisions about germplasm today has consequences that transcend the farm household. Many technical issues face today's conservers of germplasm.

Collection. In traditional agricultural systems farmers do all the collecting of germplasm. Even today in developing nations, farmers will be quick to notice unusual mutations that might serve to improve the crop in the next season. These will be carefully collected and kept for that purpose. Similarly, in traditional systems farmers will often exchange seeds with their neighbors and even with farmers in fairly distant villages. Finally, farmers will be quick to notice wild relatives of crop plants and either remove them from planted fields to the extent possible or, less often, incorporate them into new plantings. Today, crop germplasm is usually divided into categories: (1) wild relatives of crop species, (2) weedy relatives that often grow in the field with the crop plant because they are indistinguishable from it at least in their early stages of growth, (3) landraces produced over thousands of years by millions of now-forgotten farmers, (4) modern varieties produced through plant breeding, and (5) advanced breeding lines used in research institutions. Most germplasm banks focus on the first three types, though occasionally the last two are collected as well. Modern collectors have tended to sample widely, each collector bringing back (usually to the developed nations) large numbers of seed samples. Unlike farmers, however, plant collectors usually had no knowledge of the local area in which they gathered materials. Not only were they unaware of the range of materials locally available, but they were also unaware of cultural practices in either the agronomic or anthropological sense of the term. Thus even today it is rare for collectors to spend much time gathering information on planting, cultivating, and harvesting practices, on food production methods, or on local pests and diseases. Observers have noted that even passport data (i.e., the point of origin of the plant) is often missing from collections (Williams 1989). As a result, much local knowledge is lost in the collection process.

For example, a local cultivar of sorghum in West Africa may be prized for its bird resistance, its beer-making qualities, or its rapid germination. Local farmers usually know this information, but collectors from outside the region will have little or no knowledge of such attributes. Indeed, until very recently, most scientists saw local knowledge as largely irrelevant or even antithetical to the scientific enterprise.

To some extent this failure to incorporate local knowledge into scientific classifications can be overcome by the development of various laboratory tests. For example, today it is relatively easy to test baking and milling quality of wheat through various rheological and other tests performed on small samples in the lab. These tests provide information for use in the breeding process.

Yet another problem of collecting is that of sampling. An archaeologist who desires to conserve the various historical sites located in his or her country seldom has enough money to conserve all of them and must make choices about what to conserve. The same is true for germplasm. As we move from a world of peasant societies to a modern world in which far fewer people – perhaps too few (Busch and Lacy 1984a) – are located on the farm, some materials are bound to be lost.

One might think that plant explorers would thus employ well-known and established statistical sampling techniques in collecting in a given region. Yet plant explorers have rarely engaged in such systematic sampling; instead they have usually collected those materials seen as most desirable immediately in their home country. Moreover, given the differential cost of collection and the preference for paved roads and creature comforts, modern explorers have rarely made the effort to explore far from major highways (Fowler and Mooney 1990).

Classification and Evaluation. Stored materials must be cataloged in some meaningful way. As Wilkes (1984:141) suggests, "the impact of a gene depends on it being discovered and on it being valued. Values change and currently 'worthless genes' might hold the key to overcoming a vulnerability in the year 2000." Farmers have little need of complex classification systems that can be extended indefinitely across the face of the earth. They have intimate experience with the materials they use and are often able to distinguish differing characters of landraces by visual examination of the seed. At the very least, they have their own memories and the collective memories of their communities that preserve knowledge of the characters of interest locally. Therefore, a Sahelian farmer will know about the bird resistance of various millets to bird damage and will plant accordingly. In contrast, a farmer in a region where greenbug is not a sorghum pest will have no knowledge of the resistance to greenbug among the cultivars in his or her seed stocks. So much is self-evident.

In contrast, professional plant breeders and taxonomists are interested (at least in principle) not merely in the diversity found locally but in the global or regional diversity of a crop. Unlike the farmer who stores seed in a few containers around the household, breeders and taxonomists work with thousands of cultivars. They have no wish to grow all the material each year; they wish to select only a fraction of what is available for use in breeding or other research programs. No unwritten system for storage could possibly suffice.

Thus breeders and other plant scientists have tended to favor large centralized collections in which thousands of cultivars are kept in carefully labeled containers.[5] The analogy to a library is a good one here. Those of us with a few or even a few hundred books have little or no need of a universal classification system or a card catalog to help us to retrieve the volumes. In contrast, those with thousands of volumes quickly discover that classification and cataloging are essential if the books are not merely to gather dust on the shelves.

In his classic essay "The Library of Babel," Jorge Luis Borges (1962) describes a library of infinite size which contains all the books that have ever been written and all those that will ever be written in all the languages of the world.

The problem for the librarians is that there is no catalog. Moreover, the librarians disagree as to how to catalog the books. Borges makes painfully clear in this essay that a catalog must abstract certain information from that which is being cataloged. No catalog can be as complete as the information it catalogs without diminishing its own usefulness. Catalogs both provide access to the collection and restrict access to it. Each farmer in times past had intimate knowledge of a small portion of the world, but no one had access to the world as a whole. Germplasm collections will not provide access to the world as a whole either, but just to some abstracted set of entry points into the plant world.[6]

For example, the sorghum germplasm bank at the International Crops Research Institute for the Semi-Arid Tropics (ICRISAT) contains approximately 20,000 entries. When the entries are received, information about them is collected as well. The problem lies in the system of descriptors that must be used to classify the material. A review of the germplasm bank at ICRISAT revealed that no descriptors dealt with questions of storability, food quality, nonfood uses (e.g., building materials, forages), or cropping practices. Only one descriptor set dealt with grain quality. This classification system made the use of the collection to breed palatable, nutritious food more difficult than it could have been.

One potential solution to the problems of evaluating and classifying the materials in large collections is by the selection of a sample of accessions, known as a core collection. Proponents of this approach argue that core samples contain most of the variation of the collection as a whole, thereby providing a satisfactory alternative to evaluating everything in the collection (Brown 1989).

Most descriptor sets are designed for and by plant breeders. Food scientists, nutritionists, social scientists, and even pathologists and entomologists are rarely consulted in the design of descriptor sets. Hence descriptor sets necessarily limit the usefulness of germplasm banks for practical plant breeding programs for the very reasons Borges suggests. Yet at the same time, the narrow range of interests involved in the construction of the descriptor lists means that they are often even less useful than they might otherwise be.

A final problem associated with classification and evaluation involves the relative lack of communication among germplasm banks. Williams (1989) has noted that curators tend to be passive, waiting for samples and information to arrive. By not communicating with their counterparts in other nations or regions, they usually remain unaware of evaluation data created for the same cultivars at other sites. Although certain data are site-specific because of environmental differences, other data are relatively independent of the environment, making it unnecessary to replicate the evaluations in multiple locations.

Storage. Farmers store seeds for only one or two seasons in relatively simple containers. In some nations straw basketlike enclosures suffice; in others clay pots are used. These methods have stood the test of time. They may be far from the best that can be provided today, but they are usually fairly adequate to the task at hand. Moreover, because the seeds are stored for only short periods of time, the fact of storage is of little consequence for the next year's crop. And even if one farmer's seeds are lost through poor storage, the chances are that other farmers will not suffer the same loss. Thus there is little selective pressure put on the stored cultivars and there is much redundancy in the system. Moreover, there is no need to store weedy or wild relatives because these grow adjacent to or on farmers' fields. This system differs considerably from the large storage facilities currently employed.

When Vavilov established the first large-scale collection or genebank in Leningrad in the 1920s, he and his colleagues stored the seeds on shelves in a large building. It soon became apparent to curators of such collections that different types of collections were needed for different types of plants and different human purposes. Four types of collections of germplasm (usually seed) have emerged: base collections in which material is stored at low temperatures and grown out as rarely as possible, working collections used by breeders for variety development, clonal germplasm collections for vegetatively propagated crops, and genetic stock collections used for other purposes such as natural history studies. For our purposes only the first three are of relevance. Base collections are the most complete and are of particular interest to taxonomists. Working collections are of most interest to breeders because they contain material that has already been designated as promising. Clonal collections are also of interest to both taxonomists and breeders; moreover, all clonal material must be grown out regularly or it will cease to be viable.

As is explained in the following chapter, the International Plant Genetic Resources Institute (IPGRI) has recently emerged as a key actor, connected to both the Food and Agriculture Organization (FAO) and the Consultative Group for International Agricultural Research (CGIAR). Headquartered in Rome, its express purpose is to develop and coordinate an international network of germplasm banks so as to raise living standards in less developed countries. In the United States, a diffuse network of organizations is organized into the National Plant Germplasm System (NPGS) with a major (long-term) seed storage laboratory at Fort Collins, Colorado (chapter 5).

Most European nations as well as Canada and Australia also now maintain germplasm collections. In addition, India (Shrivastava et al. 1984) has significant holdings. Brazil (chapter 8), Thailand, Pakistan, China, Bangladesh, and Bulgaria have facilities planned or under construction (Plucknett et al. 1987). Finally, several private (usually voluntary) organizations also maintain genebanks (e.g., Seed Savers Exchange) (Council for Agricultural Science and Technology [CAST] 1985).

Moreover, storage at room temperature in these facilities was found to be inadequate because germination rates quickly decline for most crop plants. Thus refrigeration was soon introduced. Even today it is not at all unusual for both room-temperature and refrigerated facilities to be used for short-, medium-, and long-term storage. In each case, seeds are dried to about 5 percent moisture content. In medium-term storage facilities seeds are maintained at 1°C to 4°C, while in long-term facilities temperatures descend to –10° to –20°C. At these temperatures, many seeds will remain viable for a century or more (Towill and Roos 1989).

In addition, cryopreservation is often employed in long-term storage facilities. Cryopreservation involves the rapid freezing of seeds at temperatures of –196°C (liquid nitrogen) or from –150° to –180°C (liquid nitrogen vapor) (Towill and Roos 1989). Such low-temperature storage appears to slow the decline of germination over time.

Not all seeds can be treated with either conventional techniques or cryopreservation, however, so considerable research is under way to find the most suitable techniques for seeds or clones of such "recalcitrant" plants. Such plants include exotica but also such staples as potatoes and cassava. For these plants, which either have no seeds or have seeds that are not viable, reproduction being mainly asexual, living collections must be maintained. This is an arduous and extraordinarily expensive task. For example, Chang, Dietz, and Westwood (1989) report that the annual cost per clone for maintenance at the Corvallis, Oregon, clonal repository is $106. Moreover, such plants must be protected from pests and diseases.

Recent research has explored the possibility of using tissue culture techniques to maintain frozen samples of these and other crops. Such in vitro storage may be particularly appropriate for large seeds that are difficult to store (e.g., coconut), or when the usual means of propagation is through cuttings (e.g., citrus, cocoa, cassava) (Withers 1989). Other avenues of conservation being examined include the use of artificial seeds and the creation of DNA (deoxyribonucleic acid) libraries. With such libraries it is claimed that "the genome of all plant species of the world could easily be stored in a single small room if this, in fact, was the sensible thing to do" (Peacock 1989:371). More unlikely for the foreseeable future is the creation of DNA fragments as needed in the laboratory (Orton 1988).

Ideally, when seeds are received they are checked for disease or insect infestation. They are then cleaned and dried. If the sample is of inadequate size, as is often the case (Hanson, Williams, and Freund 1984), it is grown out to increase the sample size to that deemed minimally necessary. A small number of seeds are checked for viability. Records are kept on each accession as to the moisture content, viability, who (what other genebanks) has requested samples, and when seeds should be grown out again to ensure continued viability. At each step along the way considerable care must be taken. For example, wet seed may be damaged if frozen. Failure to keep accurate records may result in doubt as to whether the records match the samples. Failure to seal containers may result in a rise in moisture content from the air and the deterioration of the seeds (Hanson 1985). Furthermore, "Many genebanks in the tropics face rampant genetic erosion because of a lack of refrigerated storerooms, field space, trained personnel, and administrative and financial support" (Chang 1987:227).

The tasks involved are complex. For example, seed moisture content must be low to ensure conservation. If left to chance, seeds may spoil and fail to germinate at a later date. Yet predicting moisture content is difficult. In a manual for seedbanks, Hanson (1985:24) notes that "the methods outlined in the following pages can be used as a guide to prediction but experience in dealing with a crop in your own genebank will be necessary for reasonable predictions to be made." In short, a certain degree of familiarity with the seeds of a given plant – what is often called connoisseurship – is essential to successful maintenance of a seedbank.

Moreover, many of the tasks of managing a seedbank are mundane and routine. They rarely require the sophisticated equipment that is essential to a molecular biology laboratory. Nor do they produce research results that can be published in scientific journals (with the occasional exception of articles on the conservation process itself). Thus a central problem of seedbanks is that excellent scientific skills are needed but that they are employed in ways not directly rewarded by the scientific community. Of course, there are exceptions such as T. T. Chang, former director of the seedbank at IRRI, who was both a top scientist and curator. More often, however, there is a tendency to leave collection maintenance to far less skilled persons or to provide inadequate financial resources to manage the facility.

Moreover, in even the best storage facilities, seeds must be grown out after a few years to ensure that samples remain viable. This is a labor-intensive project also involving a certain level of connoisseurship (Ravetz 1971). For most closed-pollinated crops (e.g., wheat) the procedures are relatively straightforward. Small plots of each accession are grown out and the seed is harvested. All that is necessary is that the plots be properly labeled and the materials be carefully replaced in the collection. Open-pollinated crops (e.g., sorghum, maize) create more complex situations. These crops normally cross-pollinate, that is, at least two cultivars cross with each successive planting. Unlike closed-pollinated crops, open-pollinated crops are populations. If they are made to self-pollinate, the result may be considerably different from the original population. Cross-pollinations must be carefully controlled, however, if the results of the cross are to maintain the characters of the original populations. In traditional settings, farmers would normally have considerable experience in planting a few varieties so that distinct populations are maintained. Those not familiar with local conditions would have great difficulty in selecting the correct cultivars to grow together.

Furthermore, for crops that reproduce vegetatively such as potatoes and cassava, storage is considerably more difficult. Some technical research is currently under way on in vitro storage (e.g., Silva 1985), freeze-drying of pollen (Akihama and Nakajima 1978), and cryopreservation (e.g., Bajaj 1981, 1984). Even under the best of conditions, however, it appears that some variability is lost through storage.

Moreover, because all materials cannot be preserved, complex decisions must be made about what varieties to store and under what conditions they should be stored (CAST 1985). Just who should make these decisions is unclear.

One potentially overwhelming problem of management is that of eliminating duplicates within and across collections (Marshall and Brown 1981). In many if not most collections little is known about duplicated accessions. Because storage space is a scarce commodity, no curator can afford to have substantial duplication within a collection. Yet eliminating duplication requires at least some minimal evaluation, which is costly in itself. Because evaluation is an immediate cost and duplication is much less visible – especially to government officials – the all too easy solution is to ignore the problem.

Eliminating duplicates across collections is an impossible task. In the best of all possible worlds we would have only enough duplicates to ensure that accidental destruction in one bank did not result in a total loss. But political concerns about the accessibility of stored materials across national boundaries as well as a distrust of other nations' abilities to maintain materials has probably led to major duplication of materials. Moreover, the most easily available materials – the elite lines of public breeders and varieties currently grown in the fields – are those most likely to be duplicated across national lines. These are currently freely available to all breeders who request them and are well-known so breeders are likely to make those requests. The rare landraces and wild relatives are probably least likely to be found in more than one place.

Irrespective of the methods used, it should be emphasized that central storage creates simultaneously three forms of selective pressure. First, storage selects for those plants able to withstand the storage technique used. Cryopreservation will select for ability to withstand cryopreservation. Refrigeration will select for ability to withstand refrigeration, and so on. These characters may or may not be correlated with those of interest in the field or on the kitchen table. Second, storage stops selection for resistance to pests and diseases. While the seeds are stored, the pests and diseases continue to evolve. Fruit flies can make major mutations in just a few weeks. Thus when the plants are removed from storage, they may no longer exhibit the characters they exhibited upon being stored. Third, when it is necessary to grow out plants stored in collections, the conditions to which they are subjected will differ considerably from those where they were originally collected. Major crop plants appear to have originated at the so-called Vavilov centers, most of which are located in the tropical areas of the world. In those centers environmental influences on varietal change are quite different than they are in the more temperate zones where many of the germplasm banks are located. Thus the maintenance of germplasm ex situ will tend, over the long run, to change the nature of the material being stored in unknown ways. Yet another form of selective pressure will be applied to the stored varieties. No ex situ storage system can overcome these limitations because they are inherent in the very act of storage.

A logical solution to this problem would be to grow the materials in situ. Wilkes (1983:136) has argued, for example, that "these are evolutionary systems that are difficult to simulate and should not be knowingly destroyed." Reichert (1982) has even suggested that farmers be paid to maintain small acreages of rare varieties. Duvick (1977), former vice-president for research at Pioneer Hi-Bred, a large seed company, has suggested that "gene parks" (i.e., ecological preserves) be established. Others, however, have pointed to the high levels of social and political turmoil areas such as Ethiopia and the Middle East. One U.S. scientist we interviewed made an analogy with music: "I don't have any problem setting large ecosystem tracts aside. But many systems don't exist without human interaction. It is like the blues as an art form: Do you preserve the raw material for that art form by preserving poor housing and poverty?"

Still others have argued that in situ storage would be necessary and that banks would be subject to power failures.

Yet in situ maintenance may be accomplished in several ways. Most obviously, in situ conservation was what farmers did for millennia until HYVs made that practice uneconomical. It may be possible, however, for institutional structures to be developed that would provide the necessary incentives to (some) farmers to continue such practices. For example, Wood (1988) suggests that prizes be awarded at agricultural fairs based on the maintenance of landraces. Indeed, in the United States and Europe, many local landraces of garden vegetables and fruits are maintained in precisely this way without compensation and solely as labors of love. As Wood (1988:288) argues, "There are about 3 billion farming people in the world. They have almost infinite capacity, experience, and application to select and maintain crop germplasm." What is needed are means to encourage farmers to continue doing what they did well in the past. Doubtless other mechanisms as well could be developed.

Indeed, some crops simply cannot be stored in conventional germplasm collections. Maintenance of these materials can be costly, requiring significant amounts of skilled labor. Moreover, many of these crops can survive only with great difficulty outside their region of origin. Botanical gardens may play a role in this type of collection as they do in some nations. Smith (1985) has noted that botanical gardens already contain about 35,000 species, or 15 percent of all plant species, according to his estimates. Yet on a worldwide scale, few botanical gardens live up to their potential for safeguarding crop species and their relatives.

Yet another alternative is the creation of ecoparks. Under this rubric areas are set aside where wild and weedy relatives of crop plants and perhaps some varieties of the crop plants themselves can grow undisturbed by human beings. Gámez (1989) has reported some success using this approach in Costa Rica. This appears to be an excellent strategy for certain wild relatives, but it is less suitable for weedy relatives and for crop plants. The center of the problem is that crop plants have coevolved with us. They are as dependent on us as we are on them. When separated from us, more often than not, they die. In particular, as even the most casual observer of agriculture has surely noticed, crop plants grow best in soil that has been disturbed and from which competitors have been removed. In addition, the characters that make plant species survive in the wild (e.g., shattering) have often been bred out of crop plants through millennia of domestication. Similarly, weedy relatives of crop plants are often found around the edges of cultivated fields where they can take advantage of human intervention.

Only wild species are fully adapted to the conditions likely to prevail at ecoparks. Wild species often have more variability than the corresponding crop (Chapman 1989). Moreover, wild relatives may well contain characters of great significance to breeders, particularly if they are single gene characters and hence relatively easily transferred to crop plants. The history of tomato breeding, for example, has shown the importance of such wild relatives (Tigchelaar 1986). Of course, ecoparks might be modified to contain more disturbed areas that resemble cultivated fields, but this requires considerable effort. Moreover, farmers do not simply disturb the soil; they cultivate it in particular ways for particular purposes, and they have considerable knowledge about the behavior and characteristics of weedy relatives. To the extent that ecoparks do the same, their fields begin to look like those of farmers and their particular niche in the germplasm conservation system is lost.

There is also a paucity of research on in situ conservation. We know relatively little about how farmers maintain diversity and not much more about how ecoparks can be made more effective. As Williams (1989:84) puts it: "For in situ genetic reserves, little effort has yet been put into their selection and design to conserve maximum diversity, and even less has been expended on monitoring and management."

In short, there is no simple technical answer to the problems of conservation. We can no longer return to the days when the world's farmers only grew local landraces. Having taken the road to HYVs, we cannot turn back (though we might well take some detours to modes of food production that require less nonrenewable energy and are less damaging to the environment). Genebanks provide some answers but are hardly adequate substitutes for the millions of farmers who used to conserve seeds regularly. And ecoparks can contribute only in certain ways to the improvement of the overall situation. Moreover, the problem is that genetic conservation is not for a year or two but for eternity. And as Fowler and Mooney (1990:217) have noted, "Governments have never been very good at eternity."

Usability. In addition to being properly maintained and identified, plant materials must be usable by breeders. For reasons of ease as well as economic pressures, breeders prefer working with elite lines (i.e., varieties that are improved but not ready for public release). Breeders have often paid little attention to exotic materials because they are difficult to incorporate into elite lines (Kannenberg 1984). Moreover, the incorporation of exotic material into elite lines, sometimes referred to as germplasm enhancement, is not a function of many germplasm banks. Consequently, the materials are not in a form viewed as appropriate for use in research. As a result, breeders often not only are reluctant to use the exotic materials stored in these banks but view the banks as taxonomists' museums rather than research support institutions.

Availability. The problem of availability is complex. In the recent past plant genetic materials were often considered common property. Attempts by various nations to develop national control over them during the nineteenth century often met with frustration. Thus, despite legal restrictions, the British were able to obtain rubber seeds from Brazil and start an industry in Southeast Asia (Brockway 1979).

Since the 1960s and 1970s, the legal protection of plant materials has been permitted in many industrialized and a few Third World countries (see chapter 10). The term usually associated with this decision is the somewhat misleading "Plant Breeders' Rights" (PBR). Plant Variety Protection (PVP) gives certificate holders a legal monopoly over the sales of seed varieties for a given number of years, but researchers may use registered varieties to develop new cultivars and farmers may plant seed from their own harvests.

More recently, the U.S. Board of Patent Appeals has upheld utility patents on plants. The European Patent Commission is discussing legislation that would permit similar patents in member countries. Unlike PVP certificates, patents grant the holder total control over the use of the patented product for a substantial period of time, including use in future research. They also prohibit multiplication and sale by farmers. These developments, as we shall see, have had the effect of increasing the level and complexity of the controversy over germplasm use and conservation.

The issue is indeed complex. Those in the international system insist that "genetic resources are being made freely available to those who request them" (Hawkes 1985:iii), whereas others argue that the Third World countries are making political hay with this issue while ignoring the real needs of their people (Ford-Lloyd and Jackson 1984).

Conclusions

The social issues with respect to germplasm conservation are inseparable from the technical ones. Unfortunately, the agricultural research institutions of the world, with few exceptions, have had great difficulty in the past in dealing with issues that have clear normative implications. Agricultural research systems appear to a large extent to be dominated by an unsophisticated adherence to a positivistic notion of science as value-free even though agricultural research, at least in the public sector, has traditionally had a "mission" orientation, which immediately implies value commitments, at least at the institutional level. Nevertheless, individual researchers often practice their craft as if broader societally held values and value conflicts were irrelevant, both as conditions for and as consequences of their scientific work.

The resolution of these issues will require more than technical expertise. The very success of plant breeding has forced us to address our responsibilities as intelligent beings. We require a better understanding of the values that underpin current policies; we require a greater willingness on the part of all to address the different interests of the First and Third Worlds; and we require the combined efforts of technical scientists, social scientists, and policy makers to negotiate decisions that will ensure that our progeny are not denied their heritage. Our grandchildren will have to live with the decisions we make today concerning germplasm conservation. As one U.S. scientist put it, "In itself germplasm collection does nothing. But without it the imagination can't do anything." With this in mind, let us turn now to the broad international context for crop germplasm activities and policies.

3

The Global System: A Competition-Cooperation Paradox

The international system for germplasm conservation is a loosely coordinated, often politicized, network among national germplasm collections, which has encouraged the exploration, collection, evaluation, preservation, and distribution of germplasm. It is shaped in large part by the way crops are grown in the field, the techniques and national institutions available for conservation, and the political and economic issues surrounding ownership and control (i.e., rights to reproduce a particular cultivar) of germplasm in a global context. As a consequence, the system's organizations, mandates, and policies have been the focus of debates and conflict during the last two decades.

This legacy established the context for the emergence and evolution of today's international network. Nearly all observers agree that the beginnings of the contemporary international system can be traced to the brilliant and pioneering early twentieth-century work of Nikolai Vavilov, a Russian botanist. Vavilov was the first to realize the importance and potential benefits to be derived from collecting seeds of various crop cultivars from around the world and organizing them into a collection. In 1923, during his first staff meeting at the All Union Institute of Applied Botany and New Crops, Vavilov stated that the task was not only to gather plants for the immediate breeding needs of Soviet agriculture but also to save seeds from extinction (Fowler and Mooney 1990). He also recognized that modern varieties were replacing the local landraces and that plant research was destroying the very foundations of its own existence, thereby threatening global food security. Moreover, Vavilov noted that the diversity of crop plants was not evenly spread over the planet but was concentrated in a few areas. Arguing that crop diversity was the result of long-term cultivation, he hypothesized that the major food crops of the world originated in nine centers (since expanded to twelve), most of which were located in the developing nations of the world. Although recent research has suggested that current centers of diversity are not necessarily centers of origin, the concept was important for identifying sites for the most productive collection activities.

In Leningrad during the 1920s Vavilov built a germplasm collection with over 250,000 entries (Fowler and Mooney 1990), including materials collected in his travels around the world as well as by other geneticists associated with his institute. In 1925 Vavilov led Russian plant hunters on the first attempt to cover the globe in search of wild plants and primitive cultivars. These activities occurred several decades before other nations were sensitized to potential problems associated with the loss of genetic materials and to the advantages of genebanks. Although Vavilov was persecuted and died in prison during the Stalinist reign of terror, his work became the model upon which other nations have built their germplasm collections.

The FAO Crop Ecology Unit

Despite these early warnings, little attention was paid to the problem of genetic erosion internationally. The United Nations Food and Agriculture Organization (FAO) discussed the issue briefly at its founding conference in 1946. In 1961, under FAO sponsorship, the first global technical meeting was held to discuss the issues surrounding the loss of genetic materials. The emphasis of this meeting was on plant introductions and the use of wild pasture grasses. The participants were not particularly concerned with immediate action because high-yielding varieties had not yet begun to replace large numbers of landraces in developing nations (Frankel 1985).

The International Biological Programme (IBP), chaired by Otto H. Frankel, was established in 1963 by the International Council of Scientific Unions (ICSU). Among its goals was the study of plant gene pools. Frankel soon met with FAO leaders, and a division of labor was arranged in which the FAO was to focus on agricultural matters while the IBP would focus on biological issues.

It was only through the work of Erna Bennett, an Irish plant breeder, however, that genetic conservation began to receive serious international attention. Specifically, Bennett organized the second conference on genetic resources, jointly hosted by the FAO and IBP in Rome in 1967. At that time Jack R. Harlan, a crop evolution specialist at the University of Illinois, defended in situ conservation and warned of the difficulties associated with ex situ collections. The participants reached a consensus that priority would be given to preservation of primitive cultivars and landraces outside their natural place. Moreover, for the first time, the term "genetic resources" was used to describe the materials to be conserved (Frankel 1986a).

Five principles emerged from the FAO/IBP conference. The highest priority would be given to endangered landraces. The strategy taken would be generalist rather than mission-driven. Large collections would be developed. Evaluation of the collections was essential. Long-term storage procedures and facilities were to be developed. A major program was proposed to carry out the principles, but FAO had no funds for such an ambitious endeavor (Frankel 1986b). The conference did, however, mandate the establishment of a Crop Ecology Unit in 1968. The unit published a newsletter that called attention to the problem and began the first significant internationally supported collecting expeditions.

An FAO Panel of Experts on Plant Exploration and Introduction met four times between 1967 and 1974. This panel, Frankel (1986b) argued, kept the issue alive when no one else was especially interested in the subject. In 1972, Bennett and Frankel persuaded the Stockholm Conference on the Human Environment to adopt a resolution calling for international action on crop genetic conservation. For a variety of reasons, however, the international agricultural research centers (IARCs) and the Consultative Group for International Agricultural Research (CGIAR), rather than the FAO Crop Ecology Unit, were chosen to be the center of the new system. Hence, to understand the contemporary international system, it is important to trace the development of the IARCs and the CGIAR and their subsequent roles in germplasm conservation.

The International Agricultural Research Centers

During the 1940s the Rockefeller Foundation began research in Mexico in an attempt to improve Mexican agriculture. The principal focus of the program was on the improvement of wheat and hybrid corn. Similar projects emphasizing hybrid corn breeding were begun in several other Latin American countries under the auspices of the U.S. Department of Agriculture or U.S. land grant universities. Leaders of the Rockefeller Foundation had become convinced that the best way to improve the health and well-being of people was first to improve their agriculture. The emphasis on corn, however, and particularly on hybrid maize seed, produced a potential market in developing countries and the opportunity for capital accumulation for plant breeding and seed sales. Consequently, these early Rockefeller Foundation efforts were aimed at restructuring both science and society through what has come to be called the Green Revolution (Busch et al. 1991; Oasa 1981).

During the 1950s the early efforts of the Rockefeller Foundation, the U.S. government, and several foreign governments spawned a variety of agricultural programs encompassing a wide range of crops, countries, and funding agencies. In the next dozen years the international agricultural research centers were established in the Third World. The major donors were no longer the foundations but the World Bank, the U.S. Agency for International Development, and other government organizations. Both the positive and negative aspects of the Green Revolution and the role of the IARCs in it have been extensively debated (e.g., Dahlberg 1979; Kloppenburg 1988; Oasa 1981).

The role of the IARCs ranged from helping to stabilize the internal conditions of Third World countries and providing for an urban industrial class by ensuring a more secure and cheaper supply of food to preventing a "red" revolution from occurring by creating instead a "green" one (Perkins 1990). Each IARC was charged with the improvement of a particular set of crops in a particular region.

In 1971 the CGIAR was created to coordinate and extend the activities of this network of centers. Like the centers themselves, the CGIAR was created apart from the cumbersome bureaucracy and politics of FAO, thereby allowing the donors more control. Its headquarters was established within the World Bank in Washington DC.

The centers and the CGIAR have played key roles in the international plant germplasm network. Although the major goal was to increase food production, centers such as the International Maize and Wheat Improvement Center (CIMMYT) and the International Rice Research Institute (IRRI) established two of the most comprehensive world germplasm collections of wheat and maize, and rice, respectively. These centers collected, stored, and evaluated indigenous landraces and primitive cultivars that served as the raw material for the development of improved varieties and hybrids. The centers have also been vehicles for the efficient location of Third World plant genetic resources essential for developed country agriculture and the transfer of these resources to the genebanks of Europe, North America, and Japan. Significantly, these centers are located in the Vavilov centers of genetic diversity and in some important ways are the modern successors to the eighteenth- and nineteenth-century botanical gardens. Currently, nine IARCs have seed collections that are estimated to contain more than 30 percent of the world's unduplicated germplasm accessions (Raeburn 1992).

During the late 1960s and 1970s concern grew about human impacts on the environment, including genetic erosion in the Third World and genetic vulnerability in the developed nations. A consensus developed during this period on the need for a coordinated program of collection and conservation to ensure that the essential raw materials of plant improvement would not be lost. In 1972 the Technical Advisory Committee of the CGIAR sponsored a conference in

Beltsville, Maryland, to consider the possibility of creating such a network and invited all five members of the FAO panel. The Beltsville report proposed a network of nine regional genebanks coordinated by FAO. But the CGIAR did not favor decentralization and argued that its network was a more appropriate mechanism for the conservation efforts. The Technical Advisory Committee adopted a revised version of the report and agreed to support the development of a new institution. By virtue of its greater financial support, the CGIAR was able to seize the initiative and establish a new institution under its leadership.

The International Board for Plant Genetic Resources

In 1974 the major stakeholders reached a compromise that created the International Board for Plant Genetic Resources. The IBPGR was housed at the FAO Rome headquarters but operated as a CGIAR institution. The board's budget was provided not by FAO but by a group of twenty-two national governments and other member organizations of the CGIAR. With the exception of India, China, and the United Nations Environment Program, the donors represented the advanced capitalist nations. Consequently, the board's policies were set not by debate among member nations of the FAO but through decision-making processes internal to the CGIAR.

Some analysts have argued that if FAO had done its job properly there never would have been a CGIAR or IBPGR (Witt 1985). But as Jack G. Hawkes, who played a leading role in the FAO germplasm activities throughout the late 1960s and early 1970s, observed "none of our recommendations, however good, were implemented because for some reason or another FAO never seemed to find the funds. In fact, by 1971 FAO itself realized it could not fund this kind of activity, so it looked to the newly established Consultative Group to see what it could do" (quoted in Witt 1985:57).

The IBPGR's (1992:2) mandate, as stated in its 1991 annual report, was "to promote and coordinate an international network of genetic resources centres to foster the collecting, conservation, documentation, evaluation and use of plant germplasm and thereby contribute to raising the standard of living and welfare of people throughout the world." It acted as a catalyst both within and outside the CGIAR system in stimulating the action needed to sustain a viable network of institutions, organizations, and programs undertaking a range of activities to conserve plant genetic resources. Instead of attempting to follow FAO's original intention of establishing its own regional genebanks, the IBPGR functioned as a coordinator by designating existing facilities as cooperating base collections where plant germplasm materials collected by IBPGR were deposited for storage. These facilities included both national germplasm conservation units and other international centers. By the mid-1980s IBPGR reported that its network included 600 scientists working in more than 100 countries and 177 base germplasm collections in 43 genebanks. Financial support for these programs was provided by 18 nations, the UN Development Programme, and the World Bank.

One criticism of this policy of relying on existing genebanks has been that they are found principally in the industrialized North. In addition, a substantial portion of base collections located in the Third World are not national collections but rather those located in the CGIAR centers. By the late 1980s less than 15 percent of the germplasm in storage (not including that in the CGIAR centers) was located in the South (Fowler et al. 1988). Kloppenburg (1988) noted that as a consequence of this policy the developed nations, though poor in naturally occurring plant genetic diversity, are as rich in stored germplasm as the Third World. The developed nations have also been the recipients of most of the IBPGR's collection funds.

A second major activity of IBPGR has been to coordinate and fund collection activities by third parties rather than emphasizing its own expeditions. In its first ten years IBPGR's activities were focused mainly on collecting threatened germplasm with particular emphasis on crops of major economic importance and their wild and cultivated relatives. The priority crops were cereals and other staple food crops, including food legumes, oilseeds, vegetables, roots, tubers, and fruits. The IBPGR stimulated a major expansion of collection of the germplasm of these crops, often collaborating with organizations already active in the area. It mounted more than 300 collection expeditions in eighty countries yielding 120,000 new seed accessions covering 120 species. During the early years, however, most of the seed collections were for major cereal crops and high-value international crops. The absence of endangered crops of regional importance for the nutrition of poor people in these collection activities created increasing tension for the organization because Third World scientists argued that their agenda was being ignored. In part as a consequence of these concerns, IBPGR's priorities shifted during the 1980s to expand the range of crops targeted for collection.

A third significant activity of the IBPGR has been training personnel to strengthen national genetic resources programs. This was an early priority and has continued to expand during the last decade and a half. IBPGR has trained over 1600 participants in short-term technical courses and workshops although trained personnel in the field of genetic resources have remained in short supply in most developing countries. Recent efforts have included greater use of national centers and wider use of non-English programs to make the training more relevant to situations in which the skills will be applied.

One consequence of IBPGR's strategy of minimizing its own investment in infrastructure has been its reliance on existing genebanks for storage of the germplasm collected under its sponsorship and subsequent loss of control of those materials. For example, when the United States was asked by IBPGR to participate in the network of global base collections, it agreed to do so but indicated that the germplasm would become the property of the U.S. government and be made available upon request on the same basis as the rest of the collection. Although for many years the United States has generally had a policy of freely exchanging germplasm with most countries, it has allowed political considerations to dictate exclusion of a few countries (e.g., Afghanistan, Albania, Cuba, Iran, Libya). In addition to politics, some countries, including India (black pepper), Ethiopia (coffee), Ecuador (cocoa), Brazil (rubber), and Malaysia (oil palm), prohibit the exchange of certain germplasm for primarily economic reasons.

Although the IBPGR's designated global base collection has no formal legal status, the norm of free exchange has been sufficient to maintain the relatively free international flow of plant genetic materials stored in the various genebanks around the globe. Most system participants acknowledge that access has not been a serious problem to date and that on those rare occasions of political or economic denial of germplasm the informal network can generally acquire the materials through a third party. But there is no current means of enforcing the free exchange of germplasm considered by many to be a global common heritage other than by moral suasion.

During IBPGR's second decade its mandate was expanded to allow it to catalyze whatever actions were needed to sustain its international genetic resources network. This work included important elements of research and training. This change enabled IBPGR not only to fund contract work by other agencies but also to take a much more active role in coordinating efforts on a global scale and to assume a share of direct responsibility for its effectiveness.

By 1985 IBPGR's scientific and professional staff numbered 13 and its budget was $5 million. Although it remained a relatively small organization, the staff doubled and the budget increased 50 percent in the next five years. Research, which had been a secondary function, became a new priority. The sharply focused research program included methods of collecting and storage of germplasm, molecular biology, breeding systems, and techniques for sampling and measuring genetic variability. These changes increased the capacity of IBPGR to participate more actively both in the field and in research. The reduction of the collection activities was questioned, however, particularly the shift of collecting priorities to minor species important to Third World countries. In addition, because IBPGR generally has had a limited budget, a debate occurred regarding the appropriate balance between field efforts and the new research elements of the program (IBPGR 1991).

The early 1990s brought additional changes to the IBPGR. A Memorandum of Understanding on program cooperation and administration with FAO was signed and plans were initiated to establish a new institution and administration separate from FAO. The crop networks concept launched by IBPGR in 1988 continued to develop with networks for 11 crops: barley, maize, groundnut, medics, banana, sweet potato, okra, beet, rice, buckwheat, and coconut. In addition, the IBPGR-sponsored research projects, particularly for ultra-dry seed storage, pollination control methods in the regeneration of germplasm, refinement of cryopreservation techniques, and in vitro collecting and storage produced valuable results (IBPGR 1991, 1992). The IBPGR, however, remained the only international agency working solely to preserve genetic resources in agriculture and continued to be the center of major controversies between developed and Third World countries.

The International Plant Genetic Resources Institute

In 1992 in an attempt to resolve some of the controversy that had surrounded this quasi-independent UN organization, the IBPGR was transformed into a new autonomous organization called the International Plant Genetic Resources Institute (IPGRI) within the CGIAR. The decision reflected a need for a flexible and independent response to new challenges, the active involvement of many dedicated new partners, and IBPGR's conviction that national agricultural research systems should be the foundation for successful global genetic resources programs.

Geoffrey Hawtin, the director of IPGRI, noted that the increasing awareness, both politically and practically, of the importance of genetic resources conservation suggested the need for an institute that could work closely and effectively with its partners: international, national, and nongovernmental organizations, as well as scientific and development organizations (Raemond 1992). The institute endorsed a new strategic plan with four major objectives: to assist countries, particularly in the developing world, in assessing and meeting their conservation needs for plant genetic resources and strengthening their links to users; to build international collaboration in the conservation and use of plant genetic resources mainly through the encouragement of networks based on crop and geographical criteria; to work to develop and promote improved strategies and technologies for the conservation of plant genetic resources; and to provide an information service. In addition, IPGRI's regional presence was to be significantly strengthened. Five regional groups are responsible for developing and regularly reviewing IPGRI's regional strategies, providing assistance in national and regional programs, and catalyzing activities such as training, documentation, and information. Another significant change was in the composition of the board of trustees for IPGRI, which included three new board members who brought new expertise and linkages to it.

The International Undertaking

While the IBPGR was developing, changing, and eventually being transformed into the IPGRI, fierce debates were being waged at the FAO over the international control of the conservation and use of plant genetic resources. Characterized as "seed wars," these debates centered around who would collect, save, evaluate, and determine who was to have access to these resources. During the late 1970s several countries began to express concern about adequate access to genetic resources originating in the South but now stored in the North. Nonetheless, the developed nations were surprised by the resolution (No.6/81) introduced at an FAO meeting in 1981 proposing that an international system of genebanks be created under the auspices of FAO. The idea was to transfer control of germplasm from the IBPGR to FAO. The proposed change was warmly welcomed by delegates from many developing nations, from which the majority of germplasm originates, but was resisted by the nations of the North. Mooney (1983) argued that the size of the holdings in the North and the lack of them in the South meant that such holdings could be used for political purposes, as U.S. Secretary of Agriculture Earl Butz had urged a few years earlier.

Among the various concerns of the delegates to FAO was an otherwise innocuous widely circulated letter from Agricultural Research Service (ARS) administrator T. W. Edminster to Richard Demuth, chair of the IBPGR, dated 19 January 1977. That letter stated that it had been U.S. policy to exchange germplasm freely with most countries of the world, although political considerations had occasionally dictated exclusion of a few countries. Nevertheless, the United States had responded to more than 10 times as many requests for seed samples as it had made. The letter also noted that the collections at Fort Collins, Colorado, were the property of the United States government.

At the 1983 FAO meeting an additional resolution was passed (No.8/83) extending the proposal of two years earlier. An "International Undertaking on

Plant Genetic Resources'' was proposed ''to ensure that plant genetic resources of economic and/or social interest, particularly for agriculture, will be explored, preserved, evaluated and made available for plant breeding and scientific purposes. The Undertaking is based on the universally accepted principle that plant genetic resources are a heritage of mankind and consequently should be available without restriction'' (FAO 1987a:50). The Undertaking encouraged the collection, exploration, evaluation, and preservation of germplasm and aid to developing nations with storage facilities and training in plant breeding. It specifically included ''special genetic stocks (including elite and current breeders' lines and mutants)'' (FAO 1987a:50) among the materials to be collected. Moreover, an international network was to be developed ''under the auspices or the jurisdiction of FAO'' (p.52). This implied that issues involving plant breeders' rights were to be ignored and that the IBPGR was to be subordinated to FAO. The participants also passed resolution No. 9/83 instructing the FAO Council to establish a Commission on Plant Genetic Resources (CPGR) that would oversee the global system. Sixty-seven nations joined the Commission on Plant Genetic Resources at its first meeting in 1985, although several nations reserved their positions on the Undertaking, including Canada, France, the Federal Republic of Germany, Japan, Switzerland, the United Kingdom, New Zealand, and the United States.

The U.S. position focused on three issues. First, an objection was made to the inclusion of ''elite and breeders' lines'' in the FAO resolution because ''we cannot expect advanced breeding lines to be part of a public germplasm system, unless the breeder so chooses'' (USDA 1984:3) Moreover, it was argued that incorporation of these lines would make the collections unduly large. At the same time it was emphasized that even protected varieties were available for research purposes. Second, the United States was concerned that the terms ''auspices'' and ''jurisdiction'' were nowhere defined. Third, the United States supported both the IBPGR and the FAO, as evidenced by its status as a major donor. But it was felt that the IBPGR needed considerable autonomy if it was to be guided by scientific priorities; in contrast, FAO was portrayed as a ''very political organization.'' From the vantage point of the USDA, the problem was that ''the international community cannot afford to allow the IPGS [International Plant Germplasm System] to be the battleground for political or philosophical disagreements'' (USDA 1984:6).

If the views of the late William L. Brown (1988) of Pioneer Hi-Bred were representative of the U.S. seed industry, then it too was strongly opposed to the Undertaking. He attacked Pat Roy Mooney's concern for germplasm as a cover for a campaign against free enterprise. Furthermore, he asserted that "it should be recognized that plant scientists can deal much more effectively and objectively with the appropriate sharing of genetic resources than can social activists and politicians, whose understanding of the problem is minimal at best" (1988:229). The rapidity with which the issue was embraced by the Third World nations demonstrated that Mooney had touched on an issue of considerable concern and contention.

France took a position similar to that of the United States, stating that it "deplores that the Undertaking makes no mention of the Paris Convention on Plant Variety Protection, the terms of which are strictly respected by French institutions." France also urged that the relationship between FAO and the IBPGR be clarified so as to ensure the independence and flexibility of the IBPGR and that the FAO not duplicate the work of the IBPGR (FAO 1985:9).

The CPGR held its second meeting in Rome in March 1987. By then, 86 nations had joined the commission, although not all had agreed to adhere to the International Undertaking. The commission also examined the feasibility of an international fund for plant genetic resources that would emphasize the needs of developing nations. The International Organization of Consumers Unions (IOCU), represented by Mooney, proposed that a tax be levied on the seeds of improved varieties to ensure a steady income of about $150 million for the fund (*Diversity* 1987).

In addition, considerable discussion ensued over the concept of "farmers' rights" as a counterbalance to plant breeders' rights. "Most delegations which intervened on the subject stressed the importance of the concept of farmers' rights, holding that these rights derived from centuries of work by farmers which had resulted in the development of the variety of plant types which constituted the major source of plant genetic diversity" (FAO 1987b:6). As Kloppenburg (1990) has observed, this was the first time that an international body had recognized that knowledge is often produced by those outside the scientific community.

A two-year Memorandum of Understanding was signed by the FAO and the IBPGR which called for more space in FAO buildings for the IBPGR and its continued presence in Rome. Nevertheless, the United States remained the most vocal opponent among those few countries (including Canada and Japan) that did not support the Undertaking (*Diversity* 1987).

The idea of an international network was elaborated somewhat in a special report to the commission. In particular, the report emphasized that the "ultimate responsibility for the conservation of plant genetic resources . . . should rest with an intergovernmental authority such as FAO'' (FAO 1987b:2). The proposed network was seen as much broader than the limited network created by the IBPGR. Governments would be encouraged to bring their base collections under this new international jurisdiction.

The same document called for an international system of base collections, which would include wild species, landraces, and obsolete cultivars but not commercial varieties or breeders' lines. Four alternative models were proposed. In one model, resource ownership and facilities would be transferred to FAO, which would manage, administer, and be financially responsible for the collection. At the other end of the spectrum was a model that would simply require free access for breeding, research, and conservation, similar to that in existence between IBPGR and national governments. This model would leave all activities up to the individual governments, including the decision to stop conservation at some future time.

By the third session of the commission in 1989 consensus was achieved on an ''agreed interpretation'' of the Undertaking. Plant breeders' rights were to be balanced against farmers' rights. Moreover, the rights of plant breeders to protection of intellectual property were no longer viewed as incompatible with the Undertaking. This implied that the Undertaking was no longer a potential means for developing nations to gain preferential access to commercial products based on plant genetic resources (Porter 1992). Instead, it was acknowledged that all nations were donors of germplasm and users of technology. The issue of recompense for farmers was left to be resolved later. Furthermore, it was recognized that free access to germplasm did not necessarily mean access free of charge. It was agreed that the FAO would work on those crops not covered already by the IBPGR as well as in situ conservation strategies. Finally, 21 nations agreed to place their germplasm collections under the auspices of FAO.

At the fourth session, held in Rome in March 1991, the commission was able to report that ''in recent years, . . . a broad intergovernmental consensus on plant genetic resources has emerged'' (FAO 1991c:1). By then, 127 nations had become members of the commission, joined in the Undertaking, or both (FAO 1991c). In addition, 25 nations had agreed to place their collections within the FAO network although most preferred the less restrictive models. Four nations offered space for the storage of international collections. Norway offered the use of an abandoned mine where accessions could be stored in permafrost at −3.7°C (FAO 1991a).

At the same time, it was announced that a new Memorandum of Agreement between FAO and IBPGR had been signed in September 1990. It gave FAO the lead role in policy and legal matters and delegated most technical work to the IBPGR. It reaffirmed that FAO was to focus on crops complementary to those collected by IBPGR and on in situ conservation. The Memorandum also stipulated that the two organizations would jointly publish a newsletter and have representation on each other's boards. Most important, the FAO was given approval power in the selection of a new director general (FAO 1991c). The Memorandum was to remain in force for at least two years with automatic renewal unless one of the parties wished to terminate it (FAO 1991d).

The commission also worked out the outlines of a network for in situ conservation at its 1991 meeting. Of particular importance was the recognition that "the primary challenge for *in situ* conservation of genetic resources is thus not to select, set aside and guard Protected Areas containing genetic resources. Rather, it is to maintain genetic variability of the target species within a mosaic of economically and socially acceptable land use options" (FAO 1991b:2). The commission further noted that this implied a need to support local communities and participatory local institutions, including farmer organizations.

The commission celebrated its tenth anniversary at its fifth session in 1993. Numerous important topics were addressed, including the commission's role in the implementation of the landmark Convention on Biodiversity signed by more than 150 countries at the 1992 Earth Summit (United Nations Conference on Environment and Development [UNCED]) (see discussion below). The commission's secretary, José Esquinas-Alcazar opined that "the Fifth Session . . . has been the most satisfactory and far reaching one since it was established 10 years ago. During these 10 years the Commission has gone from confrontation to cooperation and we are starting to harvest the first fruits of this cooperative atmosphere" (quoted in Strauss 1993:4).[1]

Some of the most heated moments in the commission's deliberations involved the international network of ex situ base collections under the auspices or jurisdiction of FAO. One proposal offered a modified basic agreement for the CGIAR centers' collections based on the concept of trusteeship. Although the commission ultimately accepted with several modifications the proposed model as a basis for negotiations between FAO and the CGIAR centers, several delegations, and Malaysia in particular, reacted negatively to the notion of being viewed as beneficiaries of a trustee relationship to the CGIAR centers. Surprisingly, at this point, Pat Mooney, noted critic of the CGIAR policies and official of the Rural Advancement Fund International, supported the CGIAR proposal as being constructive, creative, honest, and responsible. The modifications stipulated that the commission should play a role in the development of a policy related to the CGIAR collections and that the centers develop, rather than determine, policies related to the designated germplasm. The commission then requested that the FAO director general negotiate agreements with the CGIAR centers which would be reviewed every four years by the commission (Strauss 1993).

In another key action the commission endorsed the draft International Code of Conduct for Plant Germplasm Collecting and Transfer consistent with the new Convention on Biological Diversity. A primary function of this reference document was to assist individual countries to establish their own codes or regulations until the convention was implemented. The delegates also asked the Commission Working Group to collaborate with the Commission on Sustainable Development, the governing body of the United Nations' Convention on Biological Diversity, "to develop a code relative to the implications of biotechnological development on the availability and access to plant genetic resources, genetic erosion, technology transfer, and positive or negative socioeconomic development" (Strauss 1993:5).

Finally, the commission endorsed the Working Group's support of the Fourth International Technical Conference and Programme on the Conservation and Utilization of Plant Genetic Resources scheduled to take place in Germany in 1995. This conference was expected to play a major role in implementing the Earth Summit agendas and conventions and make the global system fully operational. The commission further emphasized that the preparatory process for this conference must be participatory and country driven and it should ensure the participation of all relevant organizations and institutions dealing with ex situ and in situ conservation, as well as the sustainable use of plant genetic resources.

Despite an impressive set of accomplishments during the commission's first decade, there remained scars from the "seed wars" as the commission looked to a future fraught with difficult fiscal and political challenges. Donald N. Duvick, a member of the U.S. delegation to the commission, expressed concern over the widespread sentiment conveyed throughout the deliberations that "the only beneficiaries of plant genetic resources are the transnational seed companies." Debra Strauss, editor of *Diversity* magazine, concluded: "Thus, while the 'war' may be over, the battles are certain to continue as reinforcements are brought in by the growing number of interest groups who now have an increasing stake in the economics and politics of biodiversity" (Strauss 1993:6). The most recent of these battles was the Rio Earth Summit.

The UNCED Convention on Biological Diversity

The United Nations Conference on Environment and Development, also known as the Rio Earth Summit, marked the culmination of two and a half years of often difficult negotiations with the ambitious goal of making environmental concerns a central issue in international relations. Negotiations leading up to the summit produced five key documents. Two of the documents were legally binding: the Convention on Biological Diversity and the Climate Change Convention to curb global warming. It was hoped that three other tentative agreements would lead to binding documents sometime in the future: the *Rio Declaration,* a statement of principles to guide sustainable development; *Agenda 21,* an 800-page document serving as a blueprint for specific actions to combat a broad range of environmental problems; and a statement of principles for management, conservation, and sustainable use of tropical and temperate forests.

More than 100 heads of state and government and approximately 15,000 delegates representing 175 nations and 1500 officially accredited nongovernmental organizations (NGOs) met at the Earth Conference in June 1992 to complete formal negotiations on these five documents. At the same time, an estimated 15,000 people, drawn largely from NGOs, gathered for a parallel meeting called the Global Forum. The agenda of the Earth Conference included such problems as population growth, poverty, and sustainable development, along with more specific environmental goals such as protecting the rain forest and curbing atmospheric and oceanic pollution.

The negotiations for the Convention on Biological Diversity, begun in November 1990 under the auspices of the United Nations Environment Programme (UNEP), required six convention drafts and seven negotiating sessions. Although the political schisms did not always occur along North/South lines (Nordic countries and Australia occasionally sided with the developing nations), the negotiations did generate acute conflict between developing country and developed country interests. Developing countries were more concerned about issues of North/South equity and the use of their natural resources than they had been about similar issues in negotiations on ozone depletion or climate change. As the Malaysian delegation chief said, "Climate was theirs; biodiversity is ours" (quoted in Porter 1992:4).

The following were among the key elements of the Convention on Biological Diversity:

- Article 1, which indicated that the objectives of the convention were the conservation of biological diversity, the sustainable use of its components, and the fair and equitable share of the benefits;

- Article 15, which recognized the sovereign rights of states over their natural resources; and
- Article 16, which indicated that each contracting party recognize that technology includes biotechnology and that both access to and transfer of technology among contracting parties are essential elements for the attainment of the objectives of this convention and that access to and transfer of the technology to developing countries should be provided or facilitated under fair and most favorable terms, including concessional and preferential terms where mutually agreed.

More than 150 countries signed the Biodiversity Convention, but some issues remained unresolved. José Esquinas-Alcazar, secretary of the CPGR, indicated on the positive side that the Rio Earth Summit raised awareness of plant genetic resource issues and gave additional impetus to international efforts to manage them. It also made clear that the conservation and use of plant genetic resources could not be separated. He acknowledged, however, that this was simply a first step (Raeburn 1992).

At the opposite end of the spectrum was the reaction of the United States. Of the five UNCED documents, the Biodiversity Convention was the most controversial, in part because of the adamant refusal of President George Bush to sign the document, alone among leaders of more than 175 countries represented in Rio. Indeed, S. P. Johnson (1993:81) observed that "the refusal of the U.S. to sign the biodiversity treaty cast a long shadow over the Rio Conference, contributed to the atmosphere of North/South confrontation and weakened both U.S. and, in a wider sense, Western diplomatic efforts towards other priority objectives, such as an effective international agreement on the protection of the world's forests." Albert Gore, then one of the Senate's leading environmentalists, headed the U.S. congressional delegation to the Earth Summit and took strong exception to the rationale cited by the U.S. delegation for not signing. He charged that the Bush administration's failure to lead the effort to craft a strategy to protect both biodiversity and intellectual property rights made the Earth Summit "an economic as well as an environmental failure" (*Diversity* 1992c:23).

President Bush's decision was based on several assumptions. First, the president released a statement that the convention would lead to a loss of American jobs. The U.S. State Department also indicated that it was concerned with the convention's guidelines for the funding of biodiversity protection measures and its treatment of biotechnology and intellectual property rights. Moreover, the financing for the convention's agenda was perceived as problematic because the treaty included what some referred to as a "blank check provision." Developed countries were required to contribute money in an amount set by the parties to the convention, most of them developing countries. Great Britain and 18 other countries addressed this issue by interpreting the provision to refer only to the total amount of money needed and not the amount to be contributed by each individual nation. In addition, the financial provisions of the convention were vague, perhaps impractical, and probably could not be used to force donor countries to contribute to projects because procedures in the conference were to be decided by consensus (Porter 1992).

Intellectual property rights may well have been the key to President Bush's decision. Biotechnology trade associations, including the Industry Biotechnology Association, and the chief executives of some of the largest biotechnology companies lobbied vociferously against the treaty, arguing that it would restrict rights to intellectual property and undermine the competitive advantage of U.S. companies by forcing them to transfer valuable technology to developing countries. They further believed that it could adversely affect trade negotiations and agreements such as the General Agreement on Tariffs and Trade (GATT).

Other U.S. companies and institutions supported the treaty and suggested that the biotechnology industry's objections were based on a misinterpretation. For example, Robert Goodman, a plant pathologist at the University of Wisconsin and a former biotechnology industry executive, argued that "the U.S. biotechnology industry has jeopardized its claim to corporate responsibility for the environment. . . . It is in the biotechnology industry's best interest to seek international agreements that will assure orderly, fair, and durable arrangements for compensating those . . . who do the hard work of stabilizing and maintaining global biodiversity resources. If this important 'stewardship' of genetic resources is not accomplished, then the high profile debate over ownership and intellectual property rights could be moot" (Goodman 1992:28, 29). Other scientists noted that while the treaty included language calling for the sharing of technology covered by patents, it also indicated that such sharing should be consistent with the protection of intellectual property rights and that technology transfer should occur on "mutually agreed" terms, language included at the insistence of U.S. negotiators (Usdin 1992).

Despite efforts by India on behalf of other developing countries to undermine protection of intellectual rights in the convention, developing countries began to shift their strategy and abandon the "common heritage of mankind" position. Instead, they sought to link multinational corporations' access to genetic resources in the South to developing countries' access to products developed in the North from those genetic materials. As a consequence, they embraced the idea that access to genetic resources should be a matter for mutual agreement between countries (Porter 1992).

Finally, in the last round of negotiations the United States won a major concession in a sentence calling for "adequate and effective protection" of intellectual property rights in any technology transfer carried out under the agreement. This concession neutralized the alleged threats to U.S. business interests. The phrase "adequate and effective protection" in international trade law signifies a minimum set of norms for protection of intellectual property rights. This generally includes narrowly restricting the use of compulsory licensing, which was a primary concern of the biotechnology industry officials (Porter 1992). Ironically, several observers suggested that the United States could have better protected the interests of the biotechnology industry by signing the biodiversity agreement and participating in discussions on a protocol on international biotechnology, safety, and intellectual property rights rather than by remaining outside those discussions.

Despite the opposition expressed toward the convention by some in the developed countries, the strongest negative reactions came from those concerned about its impact on Third World countries. Germplasm activists affiliated with Genetic Resources Action International (GRAIN), a Spain-based NGO, called the convention "a small step forward for conservation" but "a large step backward for Third World control over the valuable international crop germplasm collections run by the CGIAR" (quoted in *Diversity* 1992b:5). They shared a commonly held concern about the implications of leaving existing national genebanks and the CGIAR international collections outside of the convention's jurisdiction. An international group of activists at the Earth Summit's "Global Forum" went further in drafting an "alternative bio-convention" called the *Citizen's Commitment on Biodiversity,* which called for a ban on the patenting of living things, greater control over genetic resources for farmers and indigenous peoples, and rejection of any role for the World Bank in administering funds for the conservation of biological diversity. Mooney graphically summed up the concern of many regarding the impact on developing countries:

> The Biodiversity Treaty has given itself a frontal lobotomy, ignoring the political history of the last dozen years by not using agreed-upon terms such as "farmers' rights" . . . , by not recognizing *ex situ* collections that existed prior to the treaty . . . which leaves it open to all kinds of interpretations. . . . By being too acquiescing on the issue of Intellectual Property Rights[2] (IPR), treating as a given that there will be IPR over life forms,

> and we don't take that as a given . . . , the Third World may have given up so much to have gotten so very little." (Quoted in *Diversity* 1992a:9)

Another concern regarding the convention was its potential to inadvertently lead to restrictions on the movement of germplasm, depending on the interpretation of key provisions. Geoffrey Hawtin, director of IPGRI, said that the provision describing the circumstances under which genetic resources should be provided by one nation to another could lead to extreme difficulty in getting permission to distribute materials. Yet the same provision interpreted differently could mean free and open exchange.

Henry Shands, USDA assistant deputy administrator for genetic resources and director of the USDA's National Genetic Resources Program, is concerned that developing countries are looking to the convention's Article 15, which governs access to genetic resources, as a means to receive financial support from developed countries in return for obtaining the right to use genetic resources. Developed countries in turn are apprehensive about paying for genetic materials because little is known about how to price them fairly. Moreover, if the international centers are to play a positive role, they will need to surrender some autonomy to a central authority for plant genetic resources. This will require the genetic resources units to take on more regional activities and develop better linkages with national programs (Shands 1994). Shands is also concerned that specifically agricultural issues received scant attention at Rio. Nevertheless, he has reiterated the U.S. commitment to open exchange of germplasm with all nations of the world (*Diversity* 1992b).

M. S. Swaminathan, World Food Prize Laureate and founder of the M. S. Swaminathan Research Foundation, provided a sobering summary to the Rio Earth Summit:

> The most significant accomplishment of the debate at Rio was . . . stimulating interest in making greater investments in the area of halting genetic erosion. . . . The failure was the diversion of disproportionate attention to issues relating to patenting and intellectual property rights rather than the pivotal role of genetic resources in protecting global food security and in promoting the livelihood security of the rural poor. . . . Patent protection rather than promoting a better quality of life for the poor became the major obsession. (Quoted in *Diversity* 1992a:9)

As these various perspectives suggest, the Earth Summit and the Biodiversity Convention were not panaceas but rather important next steps in global efforts to conserve the planet's biological heritage. S. P. Johnson (1993) concluded that although the Rio conference was not a triumph, it was a significant international negotiating process. The negotiating process is continuing with ever-shifting participants, rules, and objectives. Yet UNCED helped dramatically to define the terms of the debate. Perhaps the convention's greatest contribution was to outline a basis for concerted national action to conserve biological materials in a framework of international cooperation. It attempted to look beyond how to conserve the earth's genetic materials and addressed both the economic, political, and social forces and activities depleting those materials and those responsible for taking appropriate action.

The Rio Summit was followed by several United Nations actions and recommendations. In September 1992 the 47th session of the United Nations General Assembly convened in New York and endorsed actions taken at the summit. It also made recommendations on such outstanding issues as establishing a powerful new Commission on Sustainable Development, scheduling conferences proposed at the summit, and establishing a wide range of working groups and committees. The commission was to be responsible for following up on the commitments made at Rio and developing a legal and bureaucratic mechanism needed to implement the policies outlined in *Agenda 21*. The appointment of Edouard Saouma, director general of the FAO, to lead a task force to advise the UN General Assembly on the formation of a Commission on Sustainable Development led to criticisms and skepticism that the United Nations was engaged in business as usual (*Diversity* 1992e). Other observers had high hopes for the commission's future.

The UN General Assembly also moved to complete legal arrangements and establish various working committees for the two major environmental agreements signed at the summit: the Climate Change Convention and the Convention on Biological Diversity. By the time of the first anniversary of the Rio Summit (June 1993), 160 nations had signed the Convention on Biological Diversity, including the United States. The agreement, however, required ratification by the legislatures of 30 nations before it could become legally binding. Nations also needed to negotiate a series of convention protocols or specific amendments that clarified general statements in the document. The thirtieth country, Mongolia, ratified the convention on 30 September 1993. The convention entered into force on 29 December 1993.

President Bill Clinton in his first major environmental speech in April 1993 announced that the United States would sign the convention (Gallager 1993). Henry Shands, director of the U.S. National Genetic Resources Program and active participant in the numerous negotiating panels on the bioconvention, stated, "Signing of the Convention on Biological Diversity has significant implications for the National Genetic Resources Program in the United States. The nation is pledging to be a committed participant in the global conservation effort both nationally and internationally" (quoted in Gallager 1993:40). By signing the convention, Clinton demonstrated his commitment to ratification efforts which will require a three-fourths vote in the U.S. Senate. This commitment was underscored by the president's executive order creating the Council on Sustainable Development. Its 25 members will assist in drafting U.S. policies consistent with the convention's objectives of encouraging economic expansion without damaging natural resources.

The emergence of the Global Environment Facility (GEF) as the key funding mechanism to help developing countries meet the global and environmental goals agreed upon at the summit has been another important change. Established in 1990 as a three-year pilot program implemented by the UN Environment Programme, the UN Development Program, and the World Bank, the GEF was officially adopted by both the Biodiversity and Climate Conventions as their funding mechanism. Plans call for expanding the scope of GEF to include land regulation issues as they relate to the GEF's four priority global themes: loss of biodiversity, global warming, pollution of internal waters, and destruction of the ozone layer. As noted by its chair, Mohamed El-Ashry, a key challenge to the GEF is how, in the face of scarce resources, it can play "a catalytic role by integrating global and environmental considerations into the regular development assistance programs of bilateral and multi-lateral donors" (quoted in *Diversity* 1992d:6).

While the United Nations General Assembly was acting to implement the convention, the FAO Commission on Plant Genetic Resources resolved to revise the Undertaking to harmonize it with the Convention on Biodiversity by the spring of 1993. The commission called for the FAO director general "to provide a forum for negotiations among governments" for such an adaptation and "for consideration of the issue of access on mutually agreed terms to plant genetic resources, including ex situ collections, not addressed by the Convention; as well as for the issue of realization of farmers' rights" (quoted in Strauss 1993:4).

The commission further recommended that FAO should collaborate with the Secretariat of the Convention on Biological Diversity. If requested, FAO might convert the revised Undertaking into a binding legal instrument, and this might take the form of a protocol to the convention. A representative of GRAIN observed that "what was impossible ten years ago – acceptance of a legally binding agreement on plant genetic resources – might become the possible outcome of the UNCED." He further noted that "obviously the challenge is to develop a strong, equitable agreement that retains all the good elements of the Undertaking and avoids the traps of the Convention" (quoted in Strauss 1993:4).

Conclusions

The complex character of the international debate, the FAO Commission and International Undertaking, and UNCED conventions is best explained by realizing that there are no nations that are independent with respect to germplasm. As Kloppenburg and Kleinman (1988:181) noted after examining the germplasm of the 20 most important food and 20 most important industrial crops, "The general rule is substantial, and even extreme, dependence on imported genetic materials." Moreover, most transfers of plant materials are from North to North and South to South (Harlan 1988). There is little doubt, however, that the developed nations have – by virtue of their superior scientific capabilities – been able to capture much of the benefits in the worldwide transfer of germplasm from colonial times to the present.

The debate also reveals a peculiar paradox of competition and cooperation. In a capitalist world, all nations compete for resources, and each nation wishes to capture as much of the world's genetic resources as possible. At the same time, however, each nation is dependent upon many other nations for a continued supply of the desired resources. Moreover, the nature of germplasm as living, changing, dynamic entities makes it impossible for any nation to corner the market or ensure itself against all reasonable calamities.

Consequently, all nations are caught in a competition-cooperation paradox. Yet the paradox is not felt equally by all nations. Some nations, especially those in the North, are able to manipulate international decisions so that they can both compete and cooperate successfully. Others, such as Chile, are left with few choices because they have neither the requisite number of scientists to make full use of the materials nor the political and economic clout to ensure success in bi- or multilateral negotiations. Finally, a few nations, such as Brazil, stand at the edge between the developing and developed worlds. Such nations feel the competition-cooperation paradox particularly acutely because they wish both to lead the Third World and to escape from it to join the ranks of developed nations. They tend to respond by attempting to take ambiguous positions in international debates. The subsequent chapters examine how four very different nations, the United States, France, Brazil, and Chile, organize and manage their national germplasm systems in this broader international context and address the competition-cooperation paradox.

PART TWO

U.S. Perspectives

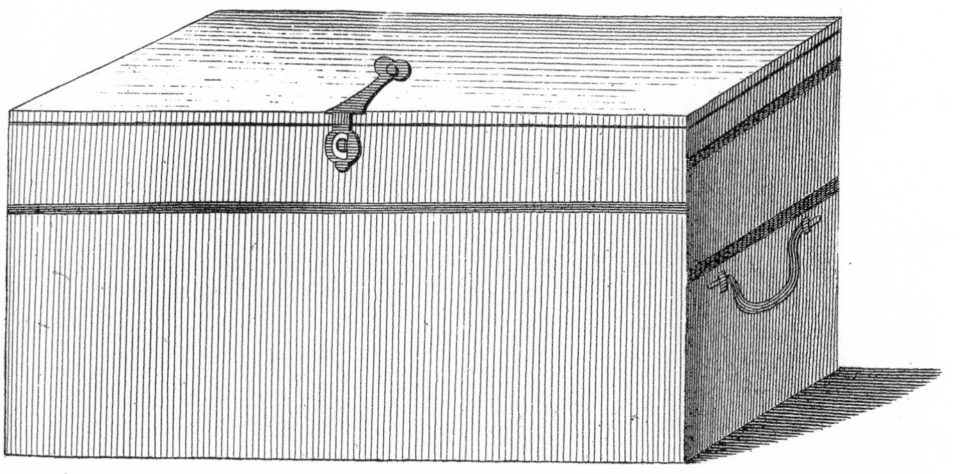

The Box with West India and W. Florida plants shut down with the openings at the ends and front left for fresh Air.

4

To Plant a Nation

God will not let us fail . . .
For . . . our work is good,
we hope to plant a nation,
Where none before hath stood.
Richard Rich, *Newes from Virginia: The Flock Triumphant,* 1610

We first knew you a feeble plant which wanted a little earth whereon to grow. We gave it to you; and afterward, when we could have trod you under our feet, we watered and protected you; and now you have grown to be a mighty tree, whose top reaches the clouds, and whose branches overspread the whole land, whilst we, who were the tall pine of the forest, have become a feeble plant and need your protection.
Red Jacket (Sagoyewatha), ca. 1792

Debates about the types of tenure systems and agriculture that the United States, a young developing nation, was to have occupied the attention of farmers after independence. Today, these same debates occupy an important place in other postcolonial regimes. One policy is based on self-sufficiency, with protective tariffs and scientific education leading to independent development strategies; the other is based on rapid integration into the world economy on the basis of free trade and export dependence and the use of science to capture markets. To generalize, we may say that the North adopted the former view while the South adopted the latter. The initial forms and later transformations of land tenure and production systems in the United States help to explain how we arrived where we are today. It is to this story that we now turn.

Colonization

European crops were introduced in North America long before permanent settlements were established. European explorers and navigators brought with them the plants necessary to make their voyages more "home-like" (Klose 1950). Colonists tried to raise these European crops when settlement began in earnest during the early seventeenth century, only to fail miserably. The first staple crop of the Americas, among both Europeans and natives, was corn, without which the settlements could not have survived (Hariot 1960). The Europeans soon differed from the indigenes in their crop practices, preferring the neat rows of homogeneous crops to the intercropping practices the natives had developed (Adair 1975; Bidwell and Falconer 1941). Thus the adaptation of "American" food crops was defined as the practice of getting them to grow under conditions of traditional European agriculture.

The creation of "neo-Europes" through the introduction of European plants, animals, and settlers was also a key component in creating the New World. By 1629, the threatening wilderness around the Jamestown settlement was converted into pastures and gardens (Crosby 1986). The Massachusetts Bay Colony imported 300 fruit trees for orchards once the initial adaptation of the land to European agriculture was finished. By 1677, it was determined that South Carolina could profitably produce rice for export; it was a net exporter of rice (primarily for the consumption of slaves in the West Indies) throughout the eighteenth century (Klose 1950).

The emerging class of gentleman farmers founded the first experimental garden for crop introduction near the Ashley River in South Carolina in 1699 (Klose 1950). Its proprietors made detailed suggestions as to which crops could be made profitable in the balmy climate. One of the crops they suggested not be grown was cotton.

Another experimental farm was established in the colony of Georgia in 1735 only to fail three years later because a suitable botanist could not be found. As the "laboratory" for scientific endeavor developed, so eventually did the scientists: John Bartram, a self-taught botanist in colonial Philadelphia, was considered by the great Swedish botanist Carolus Linneaus to be "the greatest natural botanist in the world" (Gilstrap 1961). His garden, on the banks of Philadelphia's Schuylkill River, and the Linnean Botanical Garden in Flushing, New York, were founded in 1730. The Linnean Garden was to start the never-ending introduction of European wine grapes. Charleston had two botanical gardens in 1755. These gardens were primarily for the purpose of introducing European crops into the new lands.

The work of professional botanists such as John Bartram was complemented by numerous, often self-taught amateurs, including Jane Colden, daughter of the lieutenant governor of the province of New York, Cadwallader Colden. Taught the essentials of Linnean botany by her father, she wrote a botanical tract on the flora of New York (Colden 1963) and exchanged seeds with other American botanists of the day. Jane Colden was but one of many amateurs who were fascinated by the new world around them and who helped create a uniquely North American economic botany (see also Ewan [1969] and Stuckey [1978]).

By 1750, neo-European agriculture had developed to the point that transplanted wheat was one of the two dominant crops grown in New England (along with maize). Because the wheat was genetically uniform, any threat to the crop could become an epidemic. In the early 1760s an indigenous and devastating wheat blast killed up to 30 percent of the crop.

The introduction of new plants by the founders of the new nation was coupled with the import of European lifestyles after subsistence was achieved. This is typified by the switch from corn to wheat bread as a staple, by the importation and spread of domestic animals from Europe, and by the rapid development of a European-style diet (Levenstein 1988).

Expansion

The industrious northeastern farmer and his mercantile allies in the developing craft- and merchant-oriented urban economies of New England opposed the gentleman agrarians of the South in the political battles over development policy in the new republic. The northern policy was most often associated with Alexander Hamilton, the southern with Thomas Jefferson (Monroe 1948). The outcome of this conflict is indicated by the 1787 Ordinance on the Northwest Territories.

The Northwest Ordinance provided several policies that forever differentiated the agricultural development of the United States from that of the remaining colonies of the Western Hemisphere. It denied the use of slave labor (including indentured servants) in the territory, thus protecting high-priced labor. It prevented "the perpetuation of landed estates," abolishing the European manor system and its developing Ibero-American variant (see chapters 8 and 9). It made specific provisions, based solely on population densities, for the territory to become first a series of semiautonomous territories and eventually states in the Union, equal with all the others (Carter 1934). At the same time, slavery had become rooted in the political economy of the southern states. Slavery allowed for labor-intensive cultural practices that kept profits high and capitalization low in the tobacco farms of the piedmont. Still, transportation costs, tariffs, and an eight-year war with Britain began to erode the economic viability of the slavocracy. Eli Whitney's invention of the cotton gin in 1793, however, cut production costs of cotton (Rasmussen 1974). The resulting transformation of cotton from a luxury good to something as commonplace as linen and wool extended the monoculture and plantation system in the South.

In the aftermath of the American victory over their colonial masters, the rapid push westward continued at a breakneck pace. By 1788, the Ohio Valley, suitable for extensive agriculture, was being colonized by squatters with small plots and by companies with up to one million acres (Rasmussen 1960). The squatters would arrive before the large speculators had the chance to survey their purchases. The establishment of "squatters' rights" enabled these small-holding farmers to keep or lay first claim to the lands. The speculators saw the prime Ohio Valley land as the pathway for the gradual transformation to commercial farming.

The developments of scientific agriculture were not accepted by the general farming community until the founding of the Berkshire Agricultural Society by Elkanah Watson in 1810 (Watson 1819). Modeling his practice on the agricultural fairs of Europe, Watson sponsored "show plots" and contests for the development of practical methods of scientific farming. His efforts spawned the creation of agricultural societies in New Hampshire and New York that were financed in part by their respective governments. The state Boards of Agriculture in both states were composed of the local presidents of these associations.

In 1812, John Lorain of Phillipsburg, Pennsylvania, became the first American to release an improved maize created using the scientific method of the day. Hard, indigenous flint maize of the Northeast, which matured early, was crossed with the high-yielding yellow dent variety of the south-central areas of the United States. The result was a hardy, high-yielding cultivar that was distributed throughout the new farms of the Ohio Valley (Lorain 1960 [1814]). Lorain's introduction created a flurry of activity in appraising germplasm. By 1821, at least ten different maize varieties were being bred by Euro-Americans for their color, maturation, and milling characteristics (Lorain 1975 [1814]; Bidwell and Falconer 1941).

The unique climates in the new western settlements also needed infusions of exotic germplasm to tame lands the settlers considered wilderness. Crops from the eastern part of the country had been introduced from the relatively similar western European agriculture. One way to entice farmers to move west, to populate the large tracts of open territories there, was for the government to provide the seed they needed.

Congress approved the removal of tariffs and duties on new plants introduced by the government in 1816. The plants imported from abroad were ostensibly for scientific (systematic) research, but the U.S. government also purchased bulk seed for direct distribution to farmers (Gilstrap 1961; Klose 1950).

In 1817 Watson made a private plea to the foreign consuls of the United States to send germplasm for adoption on the North American seaboard. This request led eminent botanists from around the world to send samples of seed to the United States for testing. The Spanish sent 14 varieties of wheat and one sample each of oats and barley. An 1819 Treasury Department circular encouraged United States consuls abroad to send valuable plants and seed to the American customs collectors for free distribution. President John Quincy Adams subsequently issued another circular in 1827; this time funding was provided. Henry Perrine, the consul in Campeche, Mexico, sent seeds to the treasury under the program and was given a grant of land in Florida with which to experiment with the breeding and adaptability of these plants (Rasmussen 1960). The combination of scientific knowledge and new varietal seeds was leading to commercial success, at least for publishers. Seed distribution was often an incentive to subscribe to the various agricultural journals in the new territories. In 1821, the *American Farmer* published articles suggesting the use of Chilean wheat in the territory west of the Alleghenies; included with the report was a free sample of seed (*American Farmer,* 1975 [1821]). The popularity of this approach was ensured by the high literacy rates among U.S. farmers and the fact that crop seed was the lifeblood for 90 percent of early American households (Gilstrap 1961; De Bow 1854).

Despite these importations, most of the seed used in crop production was homegrown. The seeds that would be used for new land were obtained by purchase from or barter with other farmers. One result of the expansion in area cultivated was the founding of the first commercial seedhouse, David Landreth & Son of Philadelphia, in the 1780s (Gilstrap 1961). From there the seed "industry" grew slowly. A mere 45 seed firms flourished in the East by the middle of the nineteenth century (Gilstrap 1961).

Growing crop seed was not usually a commercial enterprise in the United States, but the introduction of new crops was, and the federal government thought that subsidizing introduction was well within its interests. The priority the government put on the introduction of new crops is exemplified by the Act of 2 May 1802, entitling John James Defour to purchase four sections of land in the Ohio Territory to "promote the culture of the vine within the territory of the United States" (Klose 1950:24). The introduction of new crops became the stated purpose of agricultural science policy from then on.

Among the significant changes designed to encourage a "well-regulated" agriculture was the recommendation in 1837 by the commissioner of patents of a program of collection and free distribution of seed. Then, in the first official policy statement on food exports, Commissioner Henry L. Ellsworth suggested the export of agricultural goods to attain a more satisfactory balance of trade (Rasmussen 1960). The United States had experienced a successful foray into the world food market during the early part of the nineteenth century. The disruptions of crop production in Europe caused by the Napoleonic Wars gave eastern wheat producers an export market (Nourse 1924).

The introduction of new plants was initially a three-pronged program. The first was the procurement of novel germplasm through the Department of the Navy and the State Department's Consular Service. The navy and the merchant marine became the bearers of new varieties to America. The crew aboard Commodore Matthew C. Perry's ship on his voyage to open up the Japanese market included a patent officer charged with obtaining plants for introduction into the homeland (Klose 1950). In addition, many naval officers also owned farms and used their positions to gather plants and livestock from abroad. As holds were filled with the plants, a not-quite-contraband cargo, this practice became so disruptive of order on the navy ships that it was eventually banned.

Second, in 1839 the Agricultural Division of the Patent Office, under the control of the Department of State, was established (Klose 1950). Its plant explorers traveled the world to gather new material for testing and distribution (Klose 1950). The Agricultural Division also provided free seed, requested through a congressman, to any farmer willing to grow exotic cultivars and report the results back to the Patent Office. This created a kind of "seed patronage" system.

The growth of the federal government's free seed program was stunning. From 60,000 packages in 1847, the program grew to over 19 million packages in 1897. Appropriations for the program rose in step (Klose 1950). The year 1854 marks the first bulk purchase of "ordinary seed" from foreign countries. This practice supplanted the small scale purchase of exotic cultivars. Ironically, one of the introductions so gained was "40-Day" maize reintroduced from Spain. The government's policy of introducing plants and providing free seed was driven by the idea, held in both the public and private sectors, that "great diversities [in climate], open up a wide field of scientific investigations, to ascertain what crops and modes of culture are best adapted [for each region]" (Klose 1950:43).

By 1854, the United States was active in the multiplication of seed for distribution as well. Five acres of Washington DC was set aside on which the Patent

Office could propagate "sorgo" (i.e., sorghum) from China. These experimental plots were the direct ancestor of the thousands of acres of experimental plots that now form the national network overseen by the USDA (Gilstrap 1961).

A third means for the introduction of new plants was plant breeding. A listing of forty different bred varieties of maize was published in 1838. The method used was simple seed selection, mostly to induce early maturity of the crop and thus prevent frost damage (Bidwell and Falconer 1941).

An example is Grimm alfalfa. Wendelin Grimm, a German farmer, brought a small amount of seed alfalfa with him to Minnesota in 1858. This alfalfa was repeatedly grown out, despite the winterkill caused by the early northern frost. Grimm spent years patiently selecting new seed for its resistance to winterkill (Lyman 1960 [1903]). Grimm's alfalfa was used locally and only by chance was it "discovered" for general use in cold climates.

The external market for American farm products was being transformed as well, and the federal government was attempting to encourage farmers to orient themselves to this new market. The European demand extended to the staples of the North, beyond the traditional exports of cotton and tobacco from the slave economy in the South. Some farmers began to voice their concern over the direction government policy was headed. Many perceived the government as forcing the commercialization of production through certain crop introductions. *De Bow's Review* lamented that the 1855 report of the Bureau of Agriculture of the Patent Office was subverting the independence of the nation through the introduction of certain foreign crops. As if to concede the point that this was wrong, someone at the bureau had suggested growing opium poppies in the United States for shipment to China.

As the 1850s came to a close, the issues of slavery, industrialization, and land tenure were moving the nation toward civil war. Plant introduction was debated in terms of the advantages it gave to the freeholding developers of the West. The rapid population growth and resultant statehood of these free states was driven by the availability of crops to grow there and the restrictions, particularly in the Northwest Ordinance, on the expansion of the plantation system. Agricultural policy, like tariffs, divided North and South. Commissioner David P. Holloway of the Patent Office urged that agriculture be separated into its own department so it would be more than an appendage designed to "furnish members of Congress cutting and garden seeds to distribute among favored constituents" (quoted in Klose 1950:55).

The casual distribution of seed by the Patent Office free of charge also came under fire from the representatives of the farmers themselves. Cooperation with the Patent Office was tempered by complaints about the purity of the seed distributed. Some farmers questioned the benefits of cooperation. Yet the government depended on the farmers to whom it sent seed to collect data about the efficacy of local plant introductions. It was this research that formed the basis of initial plant introduction and evaluation. This germplasm information would in turn be used to build the science of plant breeding. A national system of seed distribution and development was needed by farmers as they became more market-oriented.

After Abraham Lincoln's election in 1860, midwestern farmers and western pioneers got the sympathetic ear they longed for; the southern slave owners did not. Secession and war followed the election and created two crucial conditions for the development of commercial agriculture in the United States. The industrialization of warfare and the number of persons involved in the fighting created an enormous internal market for food, both in the cities and in the campaigns. Lincoln also needed to solidify the antislavery constituency in the Union. One way to do so was to create more free states out of barren territories. The population potential of the Great Plains could then be used to seal the fate of slavery for future generations. Lincoln also had to discourage the western portions of the nation from leaving the Union to find their own destiny. To win the war, the Union needed western grain, railroads to carry the grain, and people to farm. The farmers needed knowledge and machinery to break the sod and transform it into cropland. Secession, by eliminating the pro-plantation congressional districts, created conditions inside and outside the government that allowed radical solutions to these problems.

The resulting flurry of legislative action during 1862 profoundly affected agriculture, plant science, and research in the United States. The Homestead Act opened far-off public lands for settlement, thereby creating demand for exotic germplasm adapted (or adaptable) to dryland, plains, and riverbed farming from North Dakota to Oklahoma, Iowa to California. The establishment of the United States Department of Agriculture solidified the advocacy of farm interests in the government with the injunction to collect and test "valuable seeds and plants," propagate those worthy, and distribute them. The Morrill Act supplied the states with funding for agricultural higher education. Finally, the Pacific Railroad Act promised access to the newly opened western lands and the subsequent access to markets needed for development.

Commercialization

During the Civil War, rising demand for foodstuffs, land, and scientific development came together to create conditions for the fastest expansion of commercial farming the world had ever seen. The new Department of Agriculture wasted no time in setting an agenda to take full advantage of the situation. Isaac Newton, the first commissioner of agriculture, set forth a six-point program for the department in 1863. It included collection of statistics on the state of agriculture in the nation; fostering the introduction of new plants and animals, both through the continuation of the free seed program and the discovery and development of new varieties; education of farmers in new cultivation techniques; testing of new agricultural implements; analysis of soils; and establishment of professorships in botany and entomology at major universities (Cochrane 1979:98).

The role of the young Department of Agriculture was to focus on scientific, not economic, intervention. The federal government was to provide access to all the knowledge available on farming and provide the technologies that made farming viable. This is considerably different from the directly economic role played by the department today. The state governments, responsible for attracting and maintaining their populations, were left to deal with the questions of economic viability (Cochrane 1979).

The war policy of the department focused on providing the nation with food and cash through expanding production and exports of wheat. Postbellum policy looked toward destroying the plantation system and introducing culture-intensive crops as a way of encouraging land reform and a more diverse diet (Klose 1950). With this policy in mind, General William T. Sherman initiated land reform in those parts of the Confederacy under his military occupation. He gave plantation lands to former slaves who had fought with the Union, just as veterans in the Mexican War had gotten scrip for western land (Murray 1946). The radical plans were later vetoed by President Andrew Johnson under the "moderate" wing of the Republican Party (Guither 1972).

The repercussions of the Civil War on agriculture in the United States extended beyond the creation of an internal urban market. The path that American farming took was the one outlined by Alexander Hamilton (1958) in his 1798 *Report on Manufactures*: industrialization would lead to capital-intensive farming.

The new states of the northern prairie went from polycultural, subsistence-based farming to monocultural commercial farming, wherein the survival of the farm as an enterprise was determined by market prices. The eastern seaboard was coming to be dominated by truck farming to supply the growing urban markets. Most of the South remained tied to a labor-intensive export economy because of weak attempts at reconstruction. Southern tenure and labor issues had to be reinterpreted with the banning of slavery, but the basic forms of accumulation remained through the development of contract farming and sharecropping (Rasmussen 1960).

Along with the concentrated scientific focus in the new USDA, Commissioner Newton brought a new seed policy. He curtailed the blanket handouts of seed and began coordinated seed placement. Only those farmers willing to use scientific methods (as then defined) would receive free seed from the government. Distribution of seed proven to be agriculturally viable was to be restricted to cases of crop failure. In 1867 the commissioner ordered that seeds be distributed only through state agricultural societies, rather than to individuals. This was the first in a slow turn of events that led to the end of free distribution of seed in 1923.

The stated desire of the department was to test up to 50,000 new plants each year (Klose 1950). The ideologues in the Department of Agriculture expected the land grant colleges to dovetail with their own science-based research and development program. This staunch belief in the efficacy of science grew in the department, reaching fruition with the appointment of James Wilson, formerly a scientist at the Iowa Agricultural Experiment Station, as the first cabinet-level secretary of agriculture in 1897 (Cochrane 1979; Klose 1950).

The attempt to reduce free seed distribution was tied to the policy of making agriculture more scientific. Congress continued its staunch support of the program, however, and international seed exchanges were accelerated. By 1871, the Department of Agriculture had completed 3450 exchanges of exotic germplasm with many of the botanical gardens of Europe and with many Asian, Ibero-American, and European governments (Klose 1950). With plant introduction, agricultural education, and scientific development policies in place, the new farmer lacked only one resource: land.

Land and land tenure were issues of critical importance in the western United States. Tenure and market domination determined the constituency for whom new plants were to be developed. In the United States, land grants were the means by which the massive stock of government lands were distributed. In 1847, the rights to hundreds of thousands of acres of land had been given away to veterans of the Mexican-American War. The three main avenues for the disposal of those public lands not given to veterans were direct sale by the government; the series of Railroad Acts, which gave away large tracts of land to be used and sold to develop the transcontinental railways; and the Homestead Act of 1862.

Direct sales to settlers by the government were usually handled under the Preemption Act of 1841, which allowed a quarter section of a township (160 acres) to be purchased by the squatting settlers before any other claims were settled. The price was $1.25 an acre for general purpose lands and $2.50 per acre for lands that were intended for the railroads. Even this price was much more than most subsistence farmers could afford. Interest rates rose to 35 percent until passage of usury laws.

Homesteaders could get their land free, provided they were over twenty-one years of age or headed a family, were or intended to become citizens, and made "improvements" on the land for five years. The Homestead Act had been proposed several times before the Civil War, each time to be defeated by a coalition of eastern business interests and slaveholders, who wanted the West to be open to speculative profits and plantation farming, respectively. In retrospect, "There would have been less speculation and more assistance given the pioneer . . . if the Homestead Act of 1862 had been passed twenty five years earlier and if at the same time the United States government had given cash instead of [land] warrants to the military veterans, railroads and educational institutions" (Murray 1946:6).

Relatively little acreage was actually opened for homesteading, although land was available for almost all those who cared to take advantage of the law. But the real developers of the Great Plains and the western frontier were the railroads. Kansas, Nebraska, and Iowa alone had a combined total of 22.1 million acres granted to the railroads compared to 3.1 million acres granted through the Homestead and Preemption Acts (Fite 1966).

Many of the potential homesteaders instead purchased railroad lands so as to have access to transportation. As one editor put it at the time, the territory from St. Paul to Fargo was "utterly unfit for the residence of man" until the railroads were built. "But no sooner does this Iron artery of trade and commerce penetrate its borders than the 'wilderness blooms like a rose'" (*Minneapolis Tribune,* quoted in Fite 1966:33). The higher price for the land was compensated for by its proximity to markets (Fite 1966). It also led to the commercialization of agriculture on these lands. This commercialization could be summed up in one word: wheat (Busch et al. 1991). The boom in the world wheat market was the ticket that led the western farmers out of the woods and onto the plains.

As the profits from western farming started to roll in during the 1870s, many investors and off-farm suppliers became increasingly interested in agriculture. Agricultural journals representing commercial producers, bankers, and suppliers, began to protest the "subsidized competition" in the form of free seed and cultural information coming from the USDA (Klose 1950). At the same time, fire sales of prime railroad land occurred when bonds collapsed during the Panic of 1873 and led to a consolidation of land that would begin the path toward bonanza farming. The pioneer families were in mortal combat with capitalists and speculators. The latter brought international attention to the huge profits these corporate farmers saw, yet failed to extract, from western land. By 1875, massive (25 percent) profits were available to the wheat farmers through the growing international and domestic markets.

The demands of market agriculture increased the need for uniform agricultural products. Land grant colleges rushed to introduce new and better wheat cultivars to satisfy this burgeoning market. The demands for uniformity and increased production sorted out all but the "best" varieties of wheat. By 1875 Turkey Red accounted for 76 percent of the Kansas wheat crop, up from 45 percent in 1870.

The epitome of monocultural, commercial, and mechanized farming practice was the bonanza farming of the Red River Valley. Landholdings in most of the Midwest were approximately 160 acres (Murray 1946). By the time of the wheat boom of the 1870s, however, the counties along the Red River from the tristate juncture of Minnesota and the Dakotas to the Canadian border had 97 farms with in excess of 10,000 acres in wheat.

Meanwhile, the ultimately more successful family farmer was eking out an existence from the sod. Diverse, low-capital operations made for long-term success. In stark contrast, "The Bonanza farms represented in agriculture many of the same characteristics and patterns found in the business world – large scale and sometimes corporate organization, absentee ownership, professional management, and specialized production – all of which were being applied in the industrial sector of the economy during the late nineteenth century" (Fite 1966:75). On one such farm in the Dakotas, up to 500 men would live in the farm dormitories during the spring, and more than 1,000 helped harvest the enormous crop.

In the South, tenure problems created by the abolition of slavery were being resolved through tenant farming and sharecropping. Sharecroppers were involved in the monocultural production of cotton dating from the Civil War. Some black farm owners actually contracted with their old masters for the production of foodstuffs (Guither 1972). The older agriculture in the industrializing eastern seaboard had completed a fundamental transformation. The farms on the outskirts of the large cities primarily provided the fresh farm goods that could not be shipped from the West (Nourse 1924).

The commercialization of agriculture also entailed the industrialization of seed production and the standardization of crops and output. Farmers became more dependent on specially bred seed to maintain the uniformity of their crops. To extend the reach of the market, the standardization of the crop had to be integrated into the larger agricultural system. As these changes took place, the federal government kept its commitment to provide free seed. Along with the distribution came research, followed by the demand for regulation, and agriculture entered a new era.

By the 1890s, the introduction of new crop plants and development of new varieties and new markets for their distribution began to change the way farmers and the government thought of seed. The new plants for U.S. farms were imported less and developed more. The breeding of varieties was eclipsing their "discovery." As plant breeding grew to national importance and acquired its own scientific rhetoric, seed quality became a social issue. Seed firms, growing the "college varieties" of crop seed, sprouted up rapidly. Many small companies started to produce and ship seeds that farmers wanted. These entrepreneurs joined the agricultural journals in casting a suspicious eye on the stolid congressional support for the free seed program. The orientation toward scientific agriculture was also beginning to impinge on the distribution program. Secretary of Agriculture Jeremiah Rusk reported in 1890 that recipients of free seed did not feel obligated to report the results of their seed testing to the department. He recommended that the bulk of future distributions be made to the experiment stations for testing (Klose 1950).

The backers of the free seed program maintained that the government should be the primary provider of seed in the nation. They pointed out that the costs of distribution were always made up through the increased value of agriculture. Their opponents argued that to distribute ordinary seed was a waste of time. The government should get out of the seed business and concentrate on the introduction of exotic germplasm. These opponents got their man in the Department of Agriculture when Julius S. Morton was appointed secretary. His 1896 edict forbidding the distribution of rare trees and flowers signaled the breakup of the Seed Division of the USDA (Klose 1950). Yet, regardless of the secretary's position, Congress was unrelentingly dedicated to the free distribution of seed.

Science: Agricultural Experiment Stations and USDA

Grover Cleveland signed the bill making the commissioner of agriculture a cabinet secretary in 1889. The secretary of agriculture became the leading advocate of science policy in the White House. His support put more bite into the Hatch Act of 1887, which authorized federal funding for state agricultural experiment stations (SAESs) at the land grant colleges. The SAESs were to provide the technical knowledge and the local adaptation plots for the introduction and development of new plants (Klose 1950).

The land grant colleges complemented the agricultural experiment stations by educating future agricultural scientists. Their rapid progress in plant breeding was joined with extensive winter lectures and other forms of adult education (Cochrane 1979).

In addition, state-sponsored county extension agents sought to extend experiment station science onto the farm through demonstration plots of new cultivars and by aiding farmers with crop suggestions for particular soils (Scott 1970). The extension agent system began on the county level and was expanded into the federal/state Cooperative Extension Service by the Smith-Lever Act of 1914. The extension services were the main vehicle for the practical scientific education of farmers. For the most part, the land grant colleges did not provide degree-related education for farmers. They trained high school teachers, extension agents, and research workers who carried out the transformation of agriculture (Cochrane 1979).

Under the direction of scientist James Wilson, the Department of Agriculture set scientific improvement of seed as one of its priorities. While the department was coordinating and scientizing the plant improvement in which farmers had engaged for years, it was also collecting germplasm from around the globe in a more systematic and comprehensive fashion. Sending plant breeders from the new experiment stations in search of suitable new germplasm and releasing these imports to the public was a key aspect of USDA seed policy. The department introduced durum wheats from the Ukraine into the Great Plains, sugar beets from west and central Europe into the black soils of the Midwest, and Bahia (navel) oranges from Brazil and Arabian dates to California (Klose 1950). By 1913, the USDA was the largest scientific concern of the federal government.

During Wilson's tenure at the Department of Agriculture the Section of Plant and Seed Introduction was established. Plant introduction became less a search and more an improvement within the new section. The rediscovery (or redefinition) of Mendel's work on inheritance and the pioneering work of Dutch botanist Hugo DeVries and American geneticist Thomas Hunt Morgan created a new theoretical basis for plant breeding and adaptation (Cochrane 1979). Justus Liebig's work on plant nutrition and that of Louis Pasteur on animal diseases raised expectations about the cornucopia to be unleashed by science in agriculture. This confidence led Seaman Knapp, who introduced rice germplasm from Japan into the South, to state that there was "no necessity for the general deterioration of farms and the too common poverty of the rural masses" (Knapp 1910:151).

The USDA's plant introduction policies led to the effective development of plant science in the United States by organizing farmers' research into a body of general knowledge and by collecting germplasm for adaptability research through the sponsorship of the experiment stations. Pooling the farmers' work was a critical task at USDA. The USDA scientists collected cultivars developed or selected by farmers and purveyed that information to the general public. As the number of varieties released grew, experiments would arise out of the trade of accessions between stations and out of the desire to develop new varieties for specific soil or climatic conditions.

Grimm alfalfa, bred cold-hardy in Minnesota, is a good example of how the networks functioned. The need for a northern alfalfa was high on USDA's priority list. In 1900, after reports of substantial and consistent alfalfa yields in the area using Grimm, the Minnesota Agricultural Experiment Station grew out the seed and released it to the public as a bona fide variety, thereby transforming the local knowledge of the region into universal science.

The case of Grimm alfalfa not only illustrates how agricultural research expanded the geographic boundaries of knowledge; it also emphasizes that the research existed without the science to codify it. Many of the new crop varieties of the late nineteenth century were developed by farmers. They were local, practical inventions, created by people who needed them to make a living. These discoveries existed whether or not they were marketable in the modern sense as seeds. They were simple applications of practical principles developed over the years to avoid certain perils of agricultural life on the frontier.

These first breeding attempts also led the way for plant breeders in the agricultural experiment stations to select and collect germplasm for their own scientific breeding work. Unfortunately, the collections that led to Garnet Chile potatoes, 40-Day maize, and Mortgage Payer wheat are, for the most part, gone (Gilstrap 1961). The Department of Agriculture used these collections as the breeding stock for its publicly funded research in crop improvement. The SAESS localized the applications of research and at the same time promoted a general, theoretical science of plant breeding.

The commitment of the federal government to scientific farming transformed the face of agriculture forever. The availability of systematically bred crop varieties, adapted to specific conditions and released free to the public, gave a massive impetus to farming. Land grant colleges provided farmers with the basic knowledge to become informed consumers of these scientific miracles. As releases were made, certain farmers were among the lucky ones to get the new foundation seed to grow out and distribute among their peers. The individual farmer receiving seed was responsible for reporting on crop yields. The experiment stations summarized these data and used them for their own reports. The farmer and the scientist wove laboratory and field into a seamless pattern of new varieties and commercial viability in the face of an ever-growing market for American foodstuffs.

By the turn of the twentieth century, the researchers began to use their scientific discoveries to develop new varieties of food crops in a laboratory environment. The federal government began to depend on the release of publicly developed seed as much as on the import of seed for the introduction of new crops (Gilstrap 1961).

Plant scientists in the experiment stations were becoming professionalized through this time. The SAES researchers began to spend more time in the laboratories and less in the field as their scientific output became valued over their role as providers of seed. The researchers themselves could not afford the time and did not have the land to grow out their new varieties. The agricultural scientific community depended on farmers' cooperation to reproduce the seed needed by American agriculture.

Farmers depended increasingly on the commercial market and large sales volumes with relatively low profit margins for their livelihoods (Nourse 1924). They required that the new varieties be guaranteed for resistance and germination, especially resistance to devastating disease. Thus, as farmers learned more about the possibility of developing new crop varieties tailored to certain conditions, they began to believe the promise that they would receive "the best nature could offer."

The United States also had begun to compete seriously in the world market. The far reach of the market and the entry of Russian wheat into the world granary demanded a set of quality standards (Nourse 1924). The grain traders and, consequently, the farmers had to deliver the uniformity demanded by various disparate and distant consumers of grain. Varietal purity and testing thus became linked with seed farming, making it a specialization.

Seed Certification

The first laboratory for testing seeds in the United States was established at the Connecticut Agricultural Experiment Station in 1876. The first USDA publication on the subject, *Rules and Apparatus for Seed Testing*, appeared in 1897 (Justice 1961). The supply and demand sides of the picture were now in place, and it only remained for some agency to arise and provide the standards needed to establish a dependable pure seed market.

At the beginning of this century, crop improvement associations and seed certification agencies began to conduct tests and disseminate information on the proper care of crops grown for seed from the public trust (Copeland and McDonald 1985). It was from this humble background that the seed certification process arose. Many of the farmers active in crop improvement were wealthy and deemed by experiment station officials to be best able to test and report on new crop varieties (Copeland, personal communication).

The political environment created by the federal Food and Drug Act of 1906 led Congress to authorize other broad regulatory powers for USDA. In 1905, USDA was directed to purchase seeds on the open market and test them for varietal and type purity. The purpose was to add credence to any labeling of seed and to create an environment of truth in labeling, distinct from the market philosophy of "let the buyer beware" (Copeland and McDonald 1985).

Testing was deemed so important to an orderly seed market that in 1908, 16 states, the USDA, and the Canadian government gathered together to form the Associations of Official Seed Analysts of North America. The organization set a goal of standard seed testing and passage of a model seed law (Justice 1961). The Seed Importation Act of 1912 marked the establishment of the first seed standards in the United States, disallowing the importation of "noxious weeds" and varietal mixes of seed (Copeland and McDonald 1985). The state was acting to set standards but left itself out of the business of achieving them.

The Wisconsin experiment station began to inspect seed grown on farms in 1913. Rogueing crews handpicked the off-variety plants out of seed farm plots to assure a harvest of pure seed. Stations in Montana, Minnesota, Missouri, and Ohio had programs to inspect the college varieties in the field by 1919. The interest of the SAESs was to guarantee the genetic purity of their own research (Parsons, Garrison, and Beeson 1961).

The seed growers' or crop improvement associations worked closely with the SAES scientists to create methods of keeping and improving seed standards. At a December 1919 meeting in Chicago, 13 mostly midwestern states founded the International Crop Improvement Association. The association agreed to encourage the improvement of crop seed and the distribution and husbandry of elite, registered, certified, and improved seeds; encourage more interest at all government levels in seed production; and assist in the standardization of seed improvement and certification work being done by member states (Copeland and McDonald 1985).

The enrollment of many actors to effect the change in seed production represented by certification was facilitated by the existence of a long tradition of government intervention in agriculture and the development of farm organizations that were waiting anxiously for a system of seed certification. In the South, the main problem was the lack of farm organizations made up of the people who worked the soil. Large landholders generally had control of farm organizations so the introduction of new techniques was dependent on the landlord system. This dependency had been interrupted by the use of extension agents to educate the southern tenants and integrate them into the knowledge system that had been created predominantly in response to northern farmers' needs (Gilstrap 1961; Scott 1970).

The major actors in the seed system linked together in the public science network were the breeders, the release committees, the foundation seed associations, and the crop improvement associations, which acted as retailers of certified seed to farmers (Copeland and McDonald 1985; Parsons, Garrison, and Beeson 1961).

The role of USDA changed in the face of this standardization. Its scientific efforts began to include detection and regulatory as well as production technology. Once the government had the technology to avoid inferior seed, its agents used both model behaviors and legislation to redefine the seed industry. The vehicle of the change was the regulatory, rather than the scientific, capacity of USDA. The final blows in this battle over the direction of the department led to the end of the free seed program in 1923, signaling both a capitulation to the private seed industry and the removal of the direct government role in plant introduction. The passage of the Federal Seed Act of 1939, based on a regulatory model, further distanced the original scientific mission of the department from actual practice.

Control of seed quality is formally independent of the federal government (Parsons, Garrison, and Beeson 1961). There are basically three types of certification agencies. Some are directly operated by the SAESs, some by state departments of agriculture, and most are quasi-independent firms closely tied administratively to the SAESs. The process described here is the most common, the certification through state-specific crop improvement associations. All crop improvement associations and most other certification agencies receive no monies from the state. Their operating costs are paid by the charges for the service they provide. Individual fees for certification of the farm, the fields, and the acreage that seed is grown on are common, and usually there is a tag fee, representing the quality testing performed on the seed after harvest. Most agencies are empowered with regulatory authority under state law.

The process has not changed formally since the original seed certification scheme (Copeland and McDonald 1985). The actors remain the same, yet their roles are defined somewhat differently because of several recent changes in agricultural practices. The issue of Plant Variety Protection (chapter 10) has intruded into the negotiations, because certain new cultivars with lucrative commercial potential are protected even though the public finances the research (Freed and Copeland, personal communications). The market path of public varieties is different from that of private varieties, in part because of the relationship to certification. State certification procedures, although somewhat standardized, have different emphases based on different philosophies of certification.

Certification is actually several certifications: varietal certification, certification of purity, and germination certification. A variety is defined in the Seed Act of 1939 as "a subdivision of a kind [i.e., a species or subspecies known by a common name, e.g., cabbage] which is characterized by growth, plant, fruit, seed, or other characteristics by which it can be differentiated from other sorts of the kind; for example, Marquis wheat, Manchu soybeans . . . and so forth" (quoted in Weiss and Little 1961). Purity concerns the presence of off-type seeds in a bag of seed purchased, especially "noxious weeds," as listed by USDA. Germination is the expected percentage of fertile seed in a bag of purchased seed.

A new variety is often developed in a breeder's laboratory, where unique characteristics are discovered and bred into a new plant. This "unnatural" selection is justified by the new cultivar's beneficial qualities for agriculture. The decision as to whether a new plant will be released as a variety is then handled by a crop committee. Most experiment stations have their own crop committees, chaired by plant breeders and made up of other scientific personnel. These committees decide if the potential new variety has characteristics that make its production as seed a benefit for agriculture. The committees are able only to release varieties for statewide use. A publicly bred variety is usually released free to the public.

In addition to the crop committee, which verifies the technical attributes of a variety, there are usually policy committees composed of a broader group of experiment station personnel, who regulate the release of varieties into the market (Copeland and McDonald 1985). This second committee determines how widely available the new variety will be and whether premiums will be paid to the breeder for its development (Copeland, personal communication).

The national criteria for verification of varietal uniqueness have been determined since 1973 by four crop-specific variety review boards. Two members each represent industry, the scientific community, and the federal government. This board reviews and evaluates information presented by breeders and advises the Association of Official Seed Certifying Agencies on the acceptability of bona fide varieties (Copeland and McDonald 1985).

The SAES crop committees are not bound by any recommendations of their national varietal review boards (Copeland and McDonald 1985). The crop committees usually accept a breeder's word about the novelty of a variety and require no additional testing. One reason for this is to avoid the appearance of conflict of interest because most crop committees are chaired by breeders developing new cultivars closely related to the ones before the committee. Another reason is a belief that the farmer should have as many choices for crop seed as possible. Though certain varieties may be better only in rare cases, to deny release would deny some farmers a viable option (Freed, personal communication).

The policy committee is less concerned with the varietal details than it is with the process of multiplying the handfuls of seed the breeder creates. To make a smooth transition from breeder to farm, releases of varieties can be targeted to create niche monopolies of seed. An example is the release of the Blackhawk bean variety in Michigan. Blackhawk had a particular disease resistance but no significant yield advantages. Thus, to release the bean openly into the public domain would have destined it to failure because no seed farmer would grow seed for such a small portion of the market. The policy committee at the Michigan Agricultural Experiment Station decided to create a monopoly in the seed. Through issuing exclusive rights, the farmer who produced the seed was guaranteed a profit based on the niche market. The seed was produced, and the farmers now have the disease-resistant bean the breeder had hoped to give them (Copeland, personal communication).

When seed for a new variety is developed, it is under the direct control of a plant breeder in a public institution or of a corporation in private breeding programs (Copeland and McDonald 1985; Parsons, Garrison, and Beeson 1961). This "breeder seed" is turned over to the regulating authority, in most cases the crop improvement association or the foundation seed growers. It becomes the seed from which the foundation seed will be grown. Four seed generations are regulated for their genetic purity and noxious contents. The first is the breeder seed generation. The second is foundation seed, which is reserved for use in continuing the variety. This is grown out as registered seed, which is preserved to propagate the crop and from which the generations of crop seed are grown. The first generation of crop seed is certified seed.

Thus certified seed ensures the farmer of high-quality seed true to varietal type and free of weeds and other debris and also guarantees uniformity. It is an assurance that all the plants of a particular variety will grow to a certain height, mature at a certain time, and be of a certain color, texture, or flavor. Moreover, that very uniformity, perhaps necessary for use in modern large-scale production systems, also has the undesirable effect of narrowing genetic diversity in the field. Therefore, the very system that is designed to shield farmers from the variation that occurs in animate nature each year carries within it the potential to subject farmers to rare but devastating crop failures by virtue of its limited genetic base.

Conclusions

As U.S. agriculture has been transformed, the diversity of crops growing in the field has been modified and usually narrowed in response to the changing demands of an increasingly industrialized society. From the introduction of new crops to the continent, to the selection of superior cultivars, to the establishment of Mendelian genetics, to the development of a seed certification system, scientists and farmers responded to the needs and desires of a changing culture. In so doing, they plowed the prairies, rearranged the ecosystem, and remade nature in their own image. At the same time, they caused crop diversity to become an issue. Eventually, their very success in socializing nature made it necessary to conserve the diversity they had eliminated. Questions about the propriety of centuries of appropriating nature – land, seeds, people's time and energy – inevitably arose. The need for a National Plant Germplasm System gradually became apparent. It is to this system that we now turn.

5

The National Plant Germplasm System: Transition and Change

> The greatest service which can be rendered any country is to add a useful plant to its culture, especially a bread grain.
>
> Thomas Jefferson

The U.S. national germplasm system has grown from its origins as an amateur occupation of colonial settlers and farmer statesmen in the 1700s to a large and diffuse network of laboratories and research stations under federal and state management. The importance of new species and varieties was recognized early in the nation's history. Hence USDA was established not to start the distribution of seeds in the United States but because seed distribution had already become a major federal activity.

Despite the enormous growth in the acquisition of foreign seeds, there was little systematic collection and cataloging until 1898, when the USDA established the Seed and Plant Introduction Section (later the Plant Introduction Office). At the same time, the Plant Introduction numbering system, still used by the National Plant Germplasm System, was established. The first plant in this system was a cabbage variety introduced from Russia in 1898 (ARS 1990).

It soon became apparent that new diseases and pests could be imported with the plants. California took the first action in 1881 to prevent distribution of the grape gall louse. Federal action occurred only in 1912, 35 years after Europe and Australia had done so, with the adoption of federal quarantine regulations (NRC 1993).

By the beginning of the twentieth century, plant introductions became the basis of crop improvement, used to breed for stress, pest resistance, and higher yields. Breeders became the curators for these introductions and maintained them as best as they could. During this time awareness about the loss of exotic plant materials arose, and greater significance was placed on germplasm management and conservation. In particular, in 1936, barley breeders Harlan and Martini (1936:315, 316) noted that the plant breeding process itself could contribute to the loss of genetic diversity. They explained: "The plant breeder has every reason to feel gratified and undoubtedly the time is not too far distant when the entire acreage will be planted to pure-line varieties. There is, however, one rather disconcerting problem raised by the plant breeder's success. In a way we lose whenever we gain. . . . The breeder is helpless without living material of diverse character."

By the 1940s the National Academy of Sciences Committee on Plant and Animal Stocks expressed concern over the fate of the materials that formed the foundation for the world's crops. In 1946 National Research Council chairman Ross Harrison, in a letter to the director general of FAO, wrote: "As a safeguard to the welfare of all peoples, steps should be taken as soon as possible to collect and maintain the plant and animal materials likely to be of service in breeding" (quoted in NRC 1991:40).

State and Federal Cooperation

Until recently, little was known about the conditions under which stored germplasm retains its ability to germinate and grow. Collections in the United States were maintained somewhat haphazardly and were highly dependent on individual plant breeders dispersed across the country. According to ARS administrator Dean Plowman, "an estimated 90 percent of the seed samples brought into this country before 1950 was lost due to inadequate knowledge and lack of storage facilities" (quoted in *Diversity* 1991:24). As a consequence of a growing concern over the loss of germplasm and the increasing importance of germplasm collections for crop improvement and increased agricultural productivity, the U.S. Congress passed the Agricultural Marketing Act in 1946. This act provided the legal basis for establishing state and federal cooperation in managing crop and livestock genetic resources and included an amendment to the Bankhead-Jones Act of 1935 to support research in this area.

The 1946 Marketing Act led to the creation of the USDA regional plant introduction stations and the National Potato Introduction Station. Before this time introductions went directly to interested researchers and breeders, and there was no requirement that they be maintained beyond their usefulness to the individual scientist. Duplicate samples were not maintained by USDA and often were lost. Since that time, however, an increasing proportion of germplasm introductions have been stored and maintained in the facilities of NPGS and are listed in the Germplasm Resources Information Network (GRIN).

The establishment of the first of four regional plant introduction stations in 1948 in Ames, Iowa, was a significant step in germplasm conservation. Three additional regional plant introduction stations were established at approximately the same time at Geneva, New York (1948); Experiment (now Griffin), Georgia (1949); and Pullman, Washington (1952). These stations were established primarily to meet the germplasm needs of plant breeders and other scientists; their responsibilities were based principally on the concerns of agriculture in each of their respective geographic regions. Their location at the site of land grant universities gave them access to a community of agricultural researchers. Finally, they provided foreign and native plant germplasm to scientists, preserved and evaluated introduced materials, and served as holding facilities for the nation's crop genetic stocks.

As the stations were established, they were incorporated into the Federal Plant Introduction System, then headquartered at the Beltsville, Maryland, Agricultural Research Center. This new system was envisioned as a cooperative enterprise that included participation by the ARS, SAESs and their associated colleges of agriculture, the Cooperative State Research Service (CSRS), and, when appropriate, the Forest Service, the Soil Conservation Service, and the Bureau of Land Management. The division of labor required a national office responsible for collecting and introducing the germplasm, as well as the four regional plant introduction stations and the other regional stations responsible for the increased maintenance, evaluation, documentation, and distribution of germplasm. Technical and administrative committees from the federal and state systems coordinated these activities.

At the regional plant introduction stations, the SAESs provided land and office space and assisted in establishing laboratories, greenhouses, and related facilities. ARS and CSRS supplied most of the funds for equipment, operating expenses, and staff. By the 1950s there was increasing recognition of the need to provide stable long-term storage that would serve as a security backup to the active collections around the country. The opening of the National Seed Storage Laboratory in Fort Collins, Colorado, in 1958 provided the principal site for long-term seed storage of genetic resources in the United States.

Other developments in the 1940s and 1950s included the Potato Introduction Project begun in 1947 by breeders to maintain valuable South American and other potato germplasm. A site for potato germplasm was established in

Sturgeon Bay, Wisconsin, in 1950. In addition, four federal plant introduction stations were active during the 1940s and 1950s. The national quarantine center and introduction station was located at Glenn Dale, Maryland. The three other stations were responsible for specialty crops.

The current National Plant Germplasm System, a diffuse network of federal, state, and private, for-profit and nonprofit institutions, agencies, and research stations (OTA 1987; NRC 1991), began to emerge in the early 1970s after a restructuring of the Agricultural Research Service. The national system has developed as an umbrella for an extensive array of germplasm management activities throughout the country. The system arose to provide better management of the germplasm of importance to U.S. agriculture and the national food supply. Germplasm activities in this system have been largely driven by an unofficial policy of national self-sufficiency that calls for comprehensive collections so as to reduce the dependence on other nations or institutions. During the past 20 years, however, the maintenance of global biological diversity by conserving the genetic diversity of crop species and their wild relatives has also become an important part of the national system.

Today, the National Plant Germplasm System, though often regarded as a well-defined entity, remains a decentralized and fragmented framework in which a multitude of individuals, committees, and USDA offices have varying levels of responsibility. Nevertheless, the system has one of the world's largest and most diverse collections of plant germplasm and has become a central player in the international network. It now shares responsibility for preserving many unique landraces no longer obtainable in their countries of origin because of habitat loss and genetic erosion. The system contains more than 350,000 accessions representing more than 8700 species. Moreover, approximately 12,000 new accessions are entered each year. At that rate an ARS study estimates that the total number of germplasm accessions in the NPGS can be expected to increase by 29 percent to 515,000 during the next 10 years (*Diversity* 1991).

The system contains virtually all of the crops of interest to U.S. agriculture. In addition, each year approximately 125,000 samples from the collections are supplied to more than 100 nations (NRC 1993). In 1992, for example, over 58,000 items were shipped from the United States to 87 countries (*Diversity* 1993d), although this represents a very small fraction of the total collection because the same varieties are requested by most nations. This pattern of requests, therefore, has the potential to erode biodiversity in the field by replacing native varieties and also concealing the decline of the more exotic materials that are in the collection but underused.

In addition, the United States plays an important formal role internationally. Eighteen specific U.S. crop collections, including those of maize, rice, sorghum, wheat, soybean, citrus, tomato, and cotton, have been designated by the IBPGR – now IPGRI – as regional or global base collections in the international network. Finally, the United States system provides backup to CGIAR collections.

Components of the System

The NPGS is composed of several stations, repositories, and laboratories with varying responsibilities and locations throughout the United States. The activities of the system include exploration, exchange, collection, and introduction; increase or regeneration; evaluation; documentation; preservation or maintenance; and distribution. Although most of the activities of the U.S. Germplasm Program take place within the NPGS, no single site is solely responsible for all of them. The following is a brief description of the components of the system and the nature of their activities:

NATIONAL SEED STORAGE LABORATORY

The National Seed Storage Laboratory (NSSL) located in Fort Collins, Colorado, is the principal site of long-term seed storage of genetic resources in the United States. The NSSL currently holds in excess of 240,000 accessions of more than 1800 species of which 60,000 are not duplicated at other sites. Germplasm in this base collection is maintained under conditions that promote long-term storage. For seeds this generally entails maintenance at low temperatures and low relative humidity and may include cryopreservation (storage in or suspended above liquid nitrogen between –150°C and –196°C) of seeds, pollen, in vitro cultures, or dormant buds, in the case of clones. It provides base collection storage facilities for the national system and holds samples of other important seed accessions. The NSSL provides backup storage for several major international collections.

As part of its comprehensive seed storage function the NSSL periodically tests seeds to ensure continued viability and, when necessary, requests that new supplies of various seeds be regenerated by regional stations, curators, and other cooperators. Viability data collected between 1979 and 1989 indicate that 29 percent of the accessions in the collection had germination rates that were either unknown or less than 65 percent. In addition, 45 percent of the accessions contained fewer than 550 seeds (NRC 1993). The NSSL does not regenerate, evaluate, enhance, or distribute germplasm. Moreover, there are no effective plans for regenerating certain seed collections held at the NSSL that are unadapted to conditions anywhere in the continental United States or Puerto Rico, even though the NPGS has accepted international responsibility for several of these base collections. In short, like other nations, the United States has emphasized storage at the expense of regeneration and evaluation. This situation highlights the need for additional facilities and the importance of cooperating with other countries in the regeneration of such germplasm.

The NSSL facility is the largest long-term storage facility in the world but recently reached its capacity. A major expansion begun in 1991 will provide additional office and laboratory space and effectively quadruple the capacity of the seed vaults. Half the vault space is expected to be used for state-of-the-art cryopreservation storage. Pilot projects are now under way. Consideration is also being given to storing cell and tissue cultures, pollen, and DNA. Although the primary mission of the NSSL is long-term storage, in its 34 years of operation it has distributed a total of more than 30,000 seed samples to plant breeders in 80 countries and engaged in research to develop improved methods for preserving seeds and plant materials.

REGIONAL STATIONS

The four regional stations noted above are responsible for maintaining the major seed-producing species held by the national system. This includes the management, regeneration, characterization, evaluation, and distribution of the seeds of more than one-third of the accessions of the national system, approximately 135,000 accessions of nearly 4,000 species. They are operated jointly by the ARS and the SAESs through the CSRS. The regional stations receive and distribute germplasm for most of the species that can be stored as dry seeds and maintain the active collections for much of the seed material in the national system. They are responsible for seed increase and for depositing backup samples in the base collections of the NSSL.

NATIONAL REPOSITORIES

The national clonal germplasm repositories contain active collections of germplasm for conserving and managing fruit, nut, and other species that cannot be held in seed collections. The eight repositories distributed over ten sites hold more than 27,000 accessions of nearly 3000 species. The primary responsibilities of the repositories are to collect, identify, propagate, preserve, evaluate, document, and distribute clonal germplasm as part of the NPGS. Long-term storage is not feasible for most material held in such collections so duplicate materials are generally maintained in field and greenhouse collections to provide some backup against loss. Many clonal crops can be conserved as seed but are impossible to maintain true to type by raising plants from seed. In addition, many clonally propagated species take a long time to mature and are best preserved as mature live plants for plant breeding and research. As a consequence, clonal collections are expensive to establish and to preserve, more labor-intensive than seed storage, and even more vulnerable than seed collections.

NATIONAL SMALL GRAINS COLLECTION

The National Small Grains Collection in Aberdeen, Idaho, is responsible for more than 110,000 accessions of wheat, barley, oats, rice, rye, *Aegilops* (a wild species related to wheat), and triticale (a hybrid of wheat and rye). Begun in 1894 as a breeder's collection, it now is the most widely used active collection in the National Plant Germplasm System. About 100,000 samples are distributed yearly from the collection to breeders, researchers, and germplasm collections in the United States and abroad.

INTERREGIONAL RESEARCH PROJECT-1

Interregional Research Project-1 (IR-1) in Sturgeon, Wisconsin, holds about 3500 potato accessions, including cultivated forms of the white or Irish potato and more than 100 related wild species. It is supported cooperatively by CSRS, ARS, and the Wisconsin Agricultural Experiment Station and is an important global resource. A variety of methods to maintain germplasm are employed; true seed potatoes, in vitro plantlets of selected clones, and tubers are maintained.

AGRICULTURAL EXPERIMENT STATIONS

The state agricultural experiment stations constitute a key component of the germplasm system. They devote significant support to plant genetics, breeding, and germplasm enhancement. For example, in 1992 the SAESs and related institutions expended more than $650 million on crop research, including genetics and crop improvement (USDA 1993). Collections at experiment stations include breeding lines, genetic stocks, landraces, and wild species related to cultivated crops. Although some of these collections are considered part of the national collections, the degree of collection duplication with the NPGS is not known. Finally, experiment stations host NPGS facilities (e.g., regional stations) and have historically provided significant in-kind support. When budgets decline or competition for funds increases, however, the in-kind contributions have been threatened and cooperative arrangements strained.

CROP COLLECTIONS

Other important crop collections of genetic resources are conserved and maintained at federal, state, and university sites. These include several crop species collections such as soybeans and other legumes, cotton, and grasses. Among these are the cotton collection in College Station, Texas (more than 5500 accessions); the long-season soybean collection in Stoneville, Mississippi (more than 3700 accessions); and the short-season soybean collection in Urbana, Illinois (nearly 10,000 accessions).

GENETIC STOCK COLLECTIONS

Genetic stock collections are accessions with unique genetic or cytological characteristics that frequently make them of particular value in basic research. These stocks differ from other germplasm stocks in many respects. Genetic stocks carry mutant genes or chromosome rearrangements, deletions, or additions. Many of these collections receive partial support from USDA through the ARS and from state resources. The genetic stock collections are important to crop development but are difficult and costly to maintain because of their unique nature and the frequently complex cytogenetics. The NPGS has made efforts to determine the location of genetic stock collections, to monitor them, and in some cases to provide funds through the ARS for maintenance. Some genetic stock centers are expanding to include the maintenance of cloned DNA sequences for use as molecular markers (restriction fragment length polymorphisms) and as specifically cloned genes. As these technologies develop and are increasingly used in research and breeding they raise new management issues and questions for the NPGS.

NATIONAL ARBORETUM

The National Arboretum in Washington DC cooperates closely with the NPGS on an informal basis. It maintains about 60,000 accessions primarily of ornamental trees and shrubs.

USDA OFFICES

In addition to the various sites, several USDA offices are responsible for data management acquisition and quarantine. The National Germplasm Resources Laboratory (NGRL) at Beltsville, Maryland, provides support for the entire national system. The Plant Introduction Office and activities related to planning and coordinating plant exploration and collection are part of the NGRL. It also maintains the Germplasm Resources Information Network, a computer database which contains information on all genetic resources preserved by the National Plant Germplasm System.

The GRIN holds the following information: (1) passport data, which includes the name of the collector, collection site data, taxonomy, and collection longitude, latitude, and elevation; (2) characterization and evaluation data, which includes general plant descriptions, agronomic responses, disease and insect pathogenic susceptibility or resistance, quality, and yield; and (3) inventory and seed requests processing data. This information serves three very different functions. The first entails obtaining and storing detailed information about accessions and requires sophisticated data-retrieval capabilities. The second is a data-classification activity for uniform and efficient data handling, and the third is inventory control to help collection managers. The problem is that in developing a database system, modifications that help one function may hinder the others.

In addition, although the structure and operation of the database management system are in place, the National Plant Germplasm System has experienced difficulties and delays in locating, correcting, and loading data that accurately represent its holdings. Accession records for much of the clonal germplasm, for example, remain unentered. The National Research Council (1991) has stated that an accurate directory or central database of all the holdings of the national system is essential and the GRIN system must better reflect the collections of the national system.

According to the National Research Council, however, the major weakness of the data management system is not the functioning of the GRIN but the paucity of data, even for listed accessions. Few records examined by the NRC committee contained useful information. The committee recommended that the NPGS obtain accurate information and ensure that accessions not presently listed be added to the database. Moreover, mechanisms for making the information held in the GRIN system more easily accessible to scientists and crop specialists must be explored.

Furthermore, the system is only as good as the data entered, and the characterization and evaluation information is generally limited to agronomic data. Information on intercropping, food quality, nutritional value, food preparation attributes, ease of digestion, and qualities of plant parts other than grain are generally lacking.

The USDA's Animal and Plant Health Inspection Service and the National Germplasm Resources Laboratory jointly manage the National Plant Germplasm Quarantine Center. This center inspects and tests plant introductions. It certifies that they are free of pests that could cause economically significant damage to U.S. crops and facilitates the movement of imported plant germplasm into the collections.

NONPROFIT ORGANIZATIONS

Finally, the national system includes informal links to several private nonprofit grass-roots organizations that often function at the margins of germplasm conservation, focus on particular types of germplasm, and play an increasingly important complementary role (chapter 6).

Administration and Leadership

Despite its important national responsibilities, the NPGS leadership and administration are scattered and often difficult to discern. The evolution of the system has produced numerous committees and individuals with varying degrees of authority and responsibility. From 1901 to 1953 the Bureau of Plant Industry in USDA had primary responsibility for germplasm exploration and collection. Since then the Agricultural Research Service has been the designated lead agency for NPGS management. It administers its programs through a decentralized system of offices and national program staff.

The national program leader for plant germplasm is charged by the ARS with planning responsibility for the national system but has little authority over budgets, programs, or management at individual sites and can only offer management recommendations. The CSRS also provides regional research funds to the NPGS as mandated by the Hatch Act. The national program leader has no authority over the distribution or use of these funds. Individual sites, such as regional plant introduction stations, are independent of each of their funding authorities, which may include ARS, CSRS, and a state agricultural experiment station. All this creates parallel and duplicate sets of authorities, responsibilities, policies, and procedures for many sites. Moreover, the national system has lacked the national and international visibility and influence needed to assure the long-term, continuing support it requires.

Providing advice for managing the national system is no less complex, and many committees and individuals hold frequently overlapping advisory responsibilities. In the recent past these have included the following:

NATIONAL PLANT GENETIC RESOURCES BOARD

The National Plant Genetic Resources Board, established in 1975, advised the secretary of agriculture and the National Association of State Universities and

Land Grant Colleges on national policy related to germplasm activities as they affect food production. Chaired by the assistant secretary of agriculture for science and education, it included public and private sector scientists. Although the National Research Council (1991) recommended greater independence and a larger advisory role on national and international policies, the board was terminated by the USDA in 1991 based on recommendations by the U.S. Office of Management and Budget and the U.S. General Services Administration (*Diversity* 1991).

NATIONAL PLANT GERMPLASM COMMITTEE

The National Plant Germplasm Committee, established in 1974, was intended to be a source of information about and an advocate for the national system and to guide and coordinate the system by developing policies, priorities, and proposals related to funding, research, and international relations. According to the 1991 NRC report, however, over the years the appointments of people who lacked direct responsibilities for the NPGS reduced its role. The NRC recommendation to disband it has been executed.

CROP ADVISORY COMMITTEES

Crop advisory committees provide expert advice on acquisition, management, and use for particular crops or crop groups (e.g., cotton, maize, wheat, tomato). Currently there are 40 committees composed primarily of agricultural scientists at land grant universities and ARS with crop-related expertise. They meet regularly and produce reports, analyses, and recommendations for the NPGS. Although the reports are not ignored, there is no mechanism for using them to set national priorities and develop plans. There is also minimal financial support.

TECHNICAL COMMITTEES

Technical committees provide technical and scientific expertise to each regional plant introduction station receiving CSRS research funds. An experiment station director serves as regional administrative adviser to each committee. Its members include a representative from each experiment station in the region and representatives from participating USDA agencies. Because the technical committees are responsible only to CSRS, there is frequently a lack of coordination and leadership among advisory groups regarding national program requirements.

TECHNICAL ADVISORY COMMITTEES

Technical advisory committees are interregional administrative advisory committees for the national clonal repositories. These committees meet separately and function independently of other advisory groups concerned with the site. They exercise no authority over programs at the site but are a source of expertise. Once again, a lack of coordination is a distinct possibility.

PLANT GERMPLASM OPERATIONS COMMITTEE

The Plant Germplasm Operations Committee, assembled by the ARS national program leader, consists of curators of the major collections, selected ARS leaders, leaders of ARS-NPGS, and representatives from the regional stations, repositories, and the National Seed Storage Laboratory. The purpose of the committee is to discuss specific questions or actions and operations between and within the national system's sites. Although it has no administrative authority, it is an ARS group that translates administrative decisions into action. It is the only NPGS activity in which the program leader exerts strong leadership and can directly affect NPGS activities.

GERMPLASM MATRIX TEAM

The Germplasm Matrix Team is chaired by the national program leader for plant germplasm and is composed of the ARS agricultural science adviser for plant germplasm and the ARS national program leaders responsible for research planning on commodities or subjects generally related to germplasm use (e.g., range, pasture and forage crops, plant health). This team provides a forum for addressing concerns about plant germplasm among program leaders who must deal with related but competing responsibilities.

The Future

This system of diverse components and complex administrative structures has come under major criticism from the U.S. General Accounting Office (GAO 1981) and the National Research Council (NRC 1991, 1993) and is undergoing significant reorganization. In the 1990 Farm Bill lawmakers established the National Genetic Resources Program to provide for the collection, preservation, and dissemination of all genetic material of importance to American food and agriculture production. It is divided into six program areas: plants, forest trees, animals, aquatics, insects, and microbes. In addition, the U.S. Congress has mandated that ARS conduct an assessment of the projected needs over the next 10 years for this new comprehensive program.

A 1992 ARS report argued that a fully implemented National Genetic Resources Program may require a quadrupling of its current $65 million annual resource commitment within the next decade (Table 5.1). Although Congress rec-

Table 5.1 Base Funding for the U.S. National Genetic Resources Program (proposed millions of 1992 dollars)

National Plant Germplasm System Program	$21.2
National Plant Genome Research Project	$16.7
National Forest Tree Genetic Resources Program	$ 0.2
National Animal Germplasm Program	$10.3
National Animal Genome Research Program	$ 9.0
National Insect Genetic Resources Program	$ 4.8
National Microbial Germplasm Program	$ 0.6
National Genetic Resources Database Support Program	$ 1.0
Total	$63.8

Source: USDA (1992). Figures represent estimates of the aggregated base funding within participating state and federal agencies, including ARS, CSRS, National Agricultural Library, Animal and Plant Health Inspection Service, Forest Service, and the SAESS.

ognized that conservation of genetic resources is essential to the future development of food production, the report observed that "only the USDA Plant Genetic Resources Program known as the NPGS, is anywhere beyond the fledgling state at this time." The report further noted that internationally coordinated plant germplasm programs, though far from complete, were more advanced than for other life forms for which they serve as a model. It completed the assessment of genetic resources by recommending that the United States "should be willing to modify its traditional policy to accept a global regime of free (open) but paid access to genetic resources, provided that access to germplasm for research and science is genuinely free" (ARS 1992).

Although the U.S. plant germplasm system is advanced relative to other genetic resource systems, the report acknowledged that there are several key projected needs for the next 10 years. The total number of germplasm accessions in the NPGS will increase dramatically if the current annual acquisition rate is maintained. The plan of work and budget for the plant program (which excludes germplasm enhancement and evaluation) is estimated to require an increase in real funding of some $10.9 million. Other recommendations included:

- Maintain full access to available plant genetic diversity of all species, acquire new material to fill gaps, and cooperate with other nations to ensure access to their collections;
- Conduct more evaluations of plant germplasm and record those data in the GRIN national database;
- Accelerate (by almost 50 percent) the rate of regenerating seed accessions to address the backlog of necessary regeneration: difficult species (some quarantinable), new acquisitions, accessions unique to the NSSL, as well as the large

number of accessions that are below the minimum desired size (550 seeds) or have lost germinability over time;

- Continue and accelerate research on difficult orthodox seeds, unorthodox seeds, and develop alternative backup storage methods for clonal collections to reduce costs and ensure a secure backup;
- Conduct, where possible, aspects of the crop germplasm survey proposed by the GAO and integrate those findings so as to establish core collections of crop species;
- Increase involvement with in situ conservation including the rare and endangered species, while recognizing that future biotechnologies may provide means to fully use almost any genetic source for gene transformations;
- Communicate information generated by the National Plant Genetic Resources Program to the national and international plant germplasm community in a more comprehensive manner than was possible previously.

Finally, the report recommended formation of a national plant germplasm coordinating committee to concern itself with policy and operations of the NPGS. This has taken on added importance because of several recent developments. The 1990 Farm Bill specifies that the National Genetic Resources Advisory Council is the primary body advising the secretary of agriculture and the director of the NGRP. This council, appointed in 1992, need include only one scientific representative for each life form represented in the program. Ironically, the formation of this new council with its limited plant germplasm representation coincided with the dissolution of both the National Plant Genetic Resources Board and the National Plant Germplasm Committee. Although it was recommended that a national coordinating committee for plant genetic resources be created to provide the needed policy and operational oversight to the NPGS, as of mid-1994 the recommendation has gone unheeded.

At this critical juncture for the newly established National Genetic Resources Program, the appointment of the director assumed increased significance. Given the controversy surrounding germplasm issues, it was encouraging that there was nearly complete consensus to appoint Henry Shands, former national plant germplasm coordinator. In addition, Shands became the first associate deputy administrator for genetic resources within the Agricultural Research Service. His appointment was applauded by government, the private sector, universities, environmentalists, and the international community alike.

The National Genetic Resources Program has received important input from the National Genetic Resources Advisory Council, which provides advice on the key issues facing the U.S. germplasm system. For example, in its spring

1993 meeting, the major topic of discussion included the impact of the Convention on Biological Diversity (see chapter 3), funding for various projects conserving genetic resources, the adequacy of various computer databases, and the public perception of biotechnology. The consensus of the council with respect to some of the possible implications of the Convention on Biological Diversity was that the United States should encourage free availability of genetic resources and should prepare contingency plans for dealing with countries that insist on commercializing their basic genetic resources.

The analysis of funding revealed continued tight budgets and no new money for germplasm conservation at the national level. Don Duvick has noted the irony in the seeming inconsistencies of U.S. germplasm policies. The government is unable or unwilling to adequately fund its genetic resources program, including its Advisory Council, while at the same time it places increasing importance on the outcome of international conferences and treaties on genetic diversity and genetic resources. The council also noted the vacuum created by the elimination of the National Plant Genetic Resources Board and the National Plant Germplasm Committee and suggested that the Agricultural Research Service stimulate the formation of "life form" committees to aid in the development and assessment of the six component programs of the National Genetic Resources Program (*Diversity* 1993b).

In addition to the reorganization within ARS, the SAES's germplasm efforts are undergoing a transition. The Experiment Station Committee on Organization and Policy has an active subcommittee on plant germplasm. Its recent agenda has included:

- The impact of utility patents on the free exchange of germplasm.
- The broad issue of utility patents for naturally occurring genes, plant traits, and breeding methodologies.
- The lack of a defined research exemption for germplasm protected via utility patents.
- The continued abuse of the farmers' exemption under the Plant Variety Protection Act.
- The continuing decline of the number of researchers in the area of applied breeding and genetics.

The committee has also reviewed and assisted in the development of the National Research Support Project (NRSP) for plant germplasm, which would replace the regional support for the four regional stations with national funding. This effort, however, has encountered much controversy and has been temporarily tabled. Finally, the subcommittee has registered concerns about the ARS implementation plan for the National Genetic Resources Program. According to the subcommittee, the ARS plan fails to recognize numerous recommendations of the National Academy of Sciences study on managing global resources.

Paralleling the specific changes regarding the U.S. National Genetic Resources Program, there has been increasing interest in the environment and natural resources in both the executive branch and Congress. Vice-President Albert Gore (1992:144) has argued that "the single most serious strategic threat to the global food system is the threat of genetic erosion." He also asserted that "the current [germplasm] system is in scandalous condition with insufficient government attention and money, little coordination between different repositories, grossly inadequate protection and maintenance of national collections, and is missing a sense of urgency where such a precious resource is concerned" (p. 140). Gore sees the problems associated with germplasm conservation as a part of the global environmental crisis, arguing that "we are, in effect, bulldozing the gardens of Eden" (p. 144).

This perspective appears to have provided a framework and foundation for a variety of activities in the administration. President Clinton established the Office on Environmental Policy to provide leadership and coordinate environmental policy in the federal government. Director Katie McGinty was previously the Senate's top staff adviser for UNCED and was instrumental in coordinating U.S. policy on the Convention on Biological Diversity (*Diversity* 1993c). Moreover, Clinton has shown his commitment to implementing the agenda of the UN Earth Summit by creating the President's Council on Sustainable Development to assist in drafting U.S. policies to fulfill obligations under the convention.

In addition, a new National Biological Survey (NBS) within the Department of Interior, has been created by Secretary Bruce Babbitt. The NBS was officially created in the fall of 1993 by assembling relevant portions of three Department of Interior bureaus: the Fish and Wildlife Service, the National Park Service, and the Bureau of Land Management. The biological survey will serve as an independent biological science bureau and build upon biological science conducted inside and outside the government. More than two-thirds of the 1994 survey budget of $180 million is dedicated to research on species biology, population dynamics, ecosystems, and inventorying and monitoring. The survey focuses on the health, abundance, and distribution of plants and animals and their ecosystems as well as conducting research necessary to understand ecological processes and their impacts on human activities (Nichols 1993).

The NBS will need to address many issues as it carries out its agenda. Some scientists believe the success of the NBS will depend in part on its ability to forge partnerships with private research entities such as the recently formed Consortium for Systematics and Biological Diversity, which includes the ARS, the American Type Culture Collection, and the National Museum of Natural History.

Another issue involves the role of the USDA's plant genetic resources system in the NBS. Essex Finney Jr., acting administrator of the ARS, has expressed concern about USDA's participation in it. He has noted the department's existing cooperation with the Interior Department and cited the National Genetic Resources Program as an example of a broad-scope program that enables research across the area of food and production agriculture for plants, insects, microbes, domestic animals, and forest trees. He has also argued that through the ARS, the Forest Service, and the Soil Conservation Service, USDA should have a role in the National Biological Survey oversight or steering committees (Nichols 1993).

Finally, Congress placed environmental issues high on its agenda for the 103d session with several specific proposals for conserving biodiversity. One House bill, the Biological Survey Act of 1993, would create a biological survey that would perform a comprehensive assessment of the U.S. biological resources, supply information to be used in protecting and managing ecosystems, and point out areas of potential conflict to the secretary of interior arising from the implementation of various laws to protect and conserve biological resources.

Several parallel congressional efforts were introduced which would establish independent centers to coordinate the nation's private and public scientific efforts in biodiversity research. One bill called for the establishment of a national center for biological resources; another would create a national center for biological diversity and conservation research within the Smithsonian Institution. This latter bill also contains provisions to establish a national biological diversity and environmental policy that would mandate preservation of biological diversity as a national goal and conservation efforts as a national priority. Finally, a House bill was introduced to establish a commission on environment and development to monitor the impact of UNCED, encourage the U.S. government and private organizations to develop programs that seek to advance the objectives of UNCED, review reports submitted to the UN Commission on Sustainable Development, and report its activities to the House and Senate periodically (Riehl 1993). These congressional and executive branch actions made biodiversity a national priority. Their positive consequences still remain to be achieved.

Conclusions

The National Plant Germplasm System is at a key juncture in its history. Several important national and international developments are having a great impact on the system. It is undergoing major internal changes, and its national and international responsibilities are being challenged. Although it is premature to evaluate the consequences of these changes and challenges, it is clear that the system will continue to evolve not as a local or regional system but as a national and international system. The NPGS will also be part of a much larger effort to conserve a wide range of germplasm. Important questions of goals and policies, management, leadership, administration, and future agendas remain. An effective way to assess and guide the system's research activities needs to be established. The management of large collections, such as those for wheat and corn, through the identification of core subsets, will need to be explored. The relative balance between in situ preservation and ex situ conservation will require constant negotiation and reassessment.

The system and its role in the global program will likely continue to grow. The NPGS will need to develop clear, concise policies and goals that encompass the conservation of plant germplasm reflecting the world's biological diversity and crop resources of immediate use to scientists and breeders. Moreover, the United States will need to continue to build cooperative programs in genetic resources with neighboring countries and other nations to conserve, collect, maintain, and regenerate germplasm. Although the NPGS has accepted responsibility for several collections in the international network, the State Department has final authority for international relations. No cohesive scientifically based policy exists to guide the nation's international activities related to plant genetic resources. Finally, as the new system emerges, provisions should be established for periodic external review of its mission, goals, personnel, programs, facilities, and future plans.

Furthermore, at the same time that the NPGS has been undergoing major change and growth, a heterogeneous and enthusiastic group of grass-roots organizations has been growing in size and importance. Although most are in their infancy, they offer the opportunity for diverse interest groups to participate as volunteers in germplasm conservation activities. Moreover, they often face the same issues confronting the national system. The next chapter provides a perspective on several of these key organizations.

6

The View from the Grass Roots

> Two key words: diversity and redundancy. These are the keys. They are not what the USDA is interested in, nor agribusiness either. They're interested in profits and economics, which are often viewed as the opposite of diversity and redundancy, but diversity and redundancy reduce risk.
>
> Researcher at an NGO

In Jessamine County, Kentucky, on the bluffs above the Kentucky River, lived a gentleman who in his lifetime experienced both the loss of a favorite vegetable variety and its rediscovery. As the story is told, when he was younger, John remembered growing and eating a variety of pumpkin that had pink flesh and an outstanding flavor. As he grew older and took a job in town to make ends meet, he stopped raising these pumpkins and lost track of them. By the time he reached retirement, they were no longer available from garden stores or seed catalogs. Then he noticed that one of his neighbors was raising this pumpkin in a garden down the road. Having carried fond memories of that pumpkin variety and its exceptional flavor for many years, he went to his neighbor and asked for a few seeds. The people along the river are good neighbors, and he received a whole jar of seed and soon began eating his favorite pumpkins once again. Today, his grandchildren continue to raise John's favorite pumpkins, both out of love for a dear relative and out of agreement with his sense of fine pumpkin flavor.

To folks like John, talk of "genetic erosion" brings to mind old fruit and vegetable varieties that have disappeared, along with the times and people they evoke; talk of "germplasm conservation" brings to mind people who save seeds. This is the view from the grass roots.

Germplasm conservation at the grass roots is a diverse and enthusiastic undertaking. Grass-roots organizations are, almost by definition, at a distance from political and economic control. Work here at the margins of germplasm conservation calls for dedication and imagination. Their guiding interests are often quite different from those of more influential government and industrial germplasm conservation institutions. But they also face many of the same dilemmas confronting major institutions and are actively devising strategies to meet these common problems.

Even at the grass roots, however, germplasm conservation is becoming predominantly a phenomenon of organizations. In the past, diversity was conserved through gardening and agriculture. Farmers and gardeners usually selected and saved their own seed, only occasionally acquiring new germplasm. With the growth of the modern seed industry and the specialization of plant breeding, however, fewer and fewer farmers are able to conserve germplasm. As this part of the structure of agriculture has changed around the world, germplasm conservation has become a separate activity from farming and from seed selection and production. Whereas once individual actions conserved germplasm diversity without any great conscious effort, in recent times germplasm conservation has become increasingly rationalized and has become the focus of specialized institutions and organizations.

In the previous chapter we examined the various government agencies responsible for germplasm conservation in the context of the development of U.S. agriculture. Yet, as in other aspects of U.S. society, grass-roots organizations are common and appear whenever an issue is defined by some segment of the public. In this chapter we move to the much smaller, less well-supported for-profit and nonprofit organizations in which private citizens do their part in germplasm conservation. Rather than providing statistics on the number of such organizations and their members – which are hard to come by – we focus on several of them for what they reveal about both seeds and people.

SSE

The Seed Savers Exchange (SSE) was established in 1975 to promote the preservation of "heirloom" vegetable varieties. In 1973, Kent Whealy, a founder and the current director of SSE, was given a pink Potato Leaf tomato, a pole bean, and a deep purple morning glory with a red star in its center. These three varieties had been passed along in the family of his wife, Diane, for at least four generations, ever since the family had left Bavaria. The varieties were given to the Whealys by Grandpa Ott. The next winter, Grandpa died.

Whealy realized that the seeds he had harvested and stored the last fall were an unusual and fragile heirloom of sorts, a family legacy. He became curious about how many other gardeners might be preserving heirloom vegetables and advertised in a few gardening publications looking for correspondents. The first year, he and six other heirloom gardeners exchanged letters and seeds. When one of them died the next year, those remaining realized that Bird Egg beans might have become extinct if not for their exchange of seed. By 1975, about 30 seed savers were corresponding, and they decided to print and circulate a list of the varieties they had to share or were looking for (Jabs 1984:47–51). Today, nearly 20 years after Whealy began seeking out other heirloom gardeners, the Seed Savers Exchange has roughly 900 members offering seed. Their yearly seed list, the *Winter Yearbook,* now has about 15,000 varieties to offer, mainly garden vegetables (Adelmann et al. 1992). Besides the *Winter Yearbook,* they also publish an annual *Harvest Edition* of news, notes, and informative articles on seed saving and germplasm conservation. In 1989 they began publishing a *Spring Edition* as well.

The philosophy of SSE is that germplasm conservation among heirloom collectors should preserve valuable varieties, true to type, and make them available to other gardeners. The early SSE thought of its conservation management system as the gardens, sheds, and root cellars of its members. The SSE practiced a form of in situ agroecosystem conservation. The venerable practice of seed saving, a part of agriculture since its beginnings but now an anachronism in Western agricultural and gardening practice, would continue to preserve these heirloom varieties as it always had. The chief hurdle to overcome was finding other interested gardeners with whom to share their seeds so as to place the heirlooms in enough locations to ensure their continued survival. In a sense, germplasm conservation for heirlooms consisted simply of finding people who would continue to grow them.

As SSE grew, so did its conservation dilemma. From the early days, when it was a network of people sharing seed of varieties they maintained for their own use, its scope and methods have grown rapidly. A substantial number of individuals came to SSE with large collections. There are people in the exchange who have amassed huge collections of squash, beans, corn, small grains, potatoes, tomatoes, and apples, among other species. Most of these collectors specialize in certain types of plants, for example, corn, vine crops, or regionally appropriate crops. Many of the varieties brought into the exchange in these collections were not heirlooms in any strict family heritage sense but simply open-pollinated varieties whose seed could be saved. The predominant motivation for many of these large collectors is a fascination with the diversity to be found in vegetable species. Jabs (1984) draws a fine picture of these large collectors in a book chapter titled "One Thousand Different Beans."

Then in 1980, when an amendment to expand the scope of the Plant Variety Protection Act (PVPA) (chapter 10) was being debated before a congressional committee, several key SSE members got involved. They were interested because of the effects they felt such legislation would have on the diversity of open-pollinated vegetable varieties commercially available to the home gardener. Their interpretation of the situation was that seed companies would drop old "standard" (open-pollinated) varieties in favor of PVP-registered varieties that would provide a higher margin of profit. At the same time, plant breeders would be concentrating their efforts on developing PVP-registered varieties that would sell in large volume to benefit from the royalties now available to them. Because the largest and most influential segment of the vegetable seed market is commercial vegetable growers, the new varieties would be designed for their needs and interests. But cultivars that appeal to commercial growers do not necessarily appeal to the home gardener. Why raise your own – for some species a time-consuming and difficult process – if the same thing can be bought cheaply in a store? One reason to garden is to raise varieties not available in the local supermarket. Therefore, the logical conclusion from this point of view was that the varieties that would soon be dropping from seed catalogs at an ever-increasing rate were "*the best home garden varieties we will ever see*" (Whealy and Adelmann 1986:7).

The concern over the extinction of open-pollinated varieties was natural for members of SSE. The heirloom varieties that had been the initial interest for the group are a special category of open-pollinated varieties. The commercial seed trade in the United States is old enough that many members have varieties in their collections as heirlooms which had been available for sale at some time in the past. Looked at from a slightly different point of view, today's commercial varieties could well become tomorrow's heirlooms. As John Withee (a SSE member, well-known in the grass-roots germplasm conservation community for his enormous collection of bean varieties) once remarked, "Even commercial beans should be treated like we do the heirlooms or other lost beans, because they sure as heck are going to be in a short while, if the seed companies decide that the sales aren't great enough" (quoted in Whealy and Adelmann 1986:127). After all, one of the primary motives of many of the members for exchanging heirlooms was to increase the diversity of available varieties with which gardeners might work, and seed catalogs have long been another source of material from which to choose. Therefore, the perceived threat to commercially available diversity raised concerns among many SSE members.

In 1981, members of SSE began a massive project to document the expected erosion of open-pollinated varieties among all commercial seed catalogs in the wake of PVPA. The initial phase of this project was completed in 1984 and published as *The Garden Seed Inventory* (Whealy 1985, 1988b, 1992). The *Inventory* brought together in one place information on what open-pollinated varieties of vegetable species were available and through which catalogs. The coverage was so thorough that the *Inventory* can be found shelved as a reference book in some university agricultural libraries, and it is also used as a reference by curators and staff of the National Plant Germplasm System.

The first *Inventory* showed that about 600 open-pollinated vegetable varieties had been lost in three years, roughly 10 percent of what was available (Whealy 1985:8). In addition, the first *Inventory* also showed that about half of what remained was available from only one or two commercial sources and in imminent danger of commercial extinction (ceasing to be available to home gardeners). Subsequent *Inventories* document the commercial demise of an additional 2200 open-pollinated varieties. Of the 5534 varieties available in 1981, 3213 varieties were still around in 1991, yielding a scant 58 percent survival rate for the decade (Whealy 1992). The proportion of varieties on the market that are commercially rare or threatened has remained in the neighborhood of 50 to 60 percent.

The *Inventories* became a tool for SSE members to use in identifying and obtaining endangered varieties worth conserving and keeping available to the public. Members are encouraged to look through the *Inventory* and purchase seed of varieties that seem endangered. This list also helps members find seed of named varieties which they are unable to obtain from local sources, acting as a concordance of seed catalogs, every gardener's January dream.

As the scope of SSE has expanded because of the enormous individual collections entering the exchange and the sharpened interest in commercial varieties, some members have seen a need to rethink and revise their approach to germplasm conservation. First, it is not clear that all the varieties being offered in the exchange are actually being conserved. Varieties may be listed in the *Winter Yearbook* one year, only to disappear the next. Members, too, may disappear from the list of those offering seeds. Is someone still keeping the varieties that are not in the current *Yearbook?* Or has a member died and the varieties he or she was maintaining become extinct? Another problem is that several members have such large collections that they cannot grow out all of their varieties in a single year. And what happens when someone with a large collection no longer feels she or he can take care of it and wants to pass most of it along?

These concerns have led to the addition of new features to SSE's system of conservation. One such addition is a Growers Network begun in 1981. The Growers Network is a core group of SSE members who have agreed to multiply selected varieties that have been identified as endangered and offer them in the *Yearbook*. Any member may volunteer to be a grower. These people not only ensure the survival of endangered varieties but also make sure that they will be available to other gardeners, thus covering both functions of germplasm conservation. The Growers Network allows SSE to pursue germplasm conservation in a less random fashion.

Another new interest of SSE is in seed storage techniques that will keep their seeds viable for more than one or two years. Storage for longer periods of time would allow SSE's relatively small active membership to take care of larger collections because they would not have to grow out the entire collection every one or two years. In investigating cold storage techniques, a conscious effort was made to develop methods suitable for home seed storage, rather than investing in equipment for use at a centralized location. A system, eventually worked out by two university researchers who have been in close contact with SSE, uses silica gel to desiccate the seeds and jars, paper envelopes, or laminated foil envelopes for storing them in a refrigerator or freezer. Another new feature is the recently acquired Heritage Farm that supplements the efforts of volunteer growers. The Heritage Farm is something of a paradox for SSE. The farm contradicts the strategy of diffused conservation, but it also serves as a showcase for what they are trying to do. SSE has found that the sight of hundreds of diverse varieties of the same vegetable species growing in one location gets an enormous positive reaction. The Heritage Farm thus serves as much as an educational tool as it does as a conservator of plant varieties.

The interests and activities of the Seed Savers Exchange have expanded beyond the original ones, but the SSE has made a conscious effort to maintain its original focus. Heirloom varieties have not been neglected as activities have expanded to include all endangered open-pollinated varieties. Storage and growing out activities have encouraged centralization, but SSE has managed to retain most activities in the hands of its member-gardeners through the *Winter Yearbook*. The *Yearbook* remains the most basic effort and is a forum of germplasm and information exchange among a diffuse membership. The SSE has promoted the formation of other groups and was the model for several of them, including the (now defunct) Grain Exchange and Seeds Blum.

NAFEX

The North American Fruit Explorers (NAFEX) trace their organizational roots back as far as any grass-roots organization involved in germplasm conservation. NAFEX gradually evolved out of the Northern Nut Growers Association, a group of people interested in finding, propagating, and improving nut-bearing trees. The Northern Nut Growers Association has been in existence for about 85 years and may be the oldest grass-roots organization that collects germplasm still in existence in the United States. Several of the founders of NAFEX first met through a persimmon interest group within the Northern Nut Growers. Although the persimmon is a fruit, propagating it requires skills similar to that for growing nut trees (Fishman 1986:17). This group of persimmon growers shared an interest in improving and breeding persimmons and in extending the range in which persimmons are grown to climatically marginal regions.

Members of the persimmon interest group corresponded and exchanged scionwood. In the 1950s this group evolved into a round-robin letter exchange, and their interests expanded to include other species beyond persimmons. By the early 1960s there were eight regional groups within a network called the Round Robin of Corresponding and Contributing Horticultural Hobbyists (Fishman 1986:14). These groups continued to grow until they became too ponderous, and the gathered letters were not making it all the way around the circuit. Milo Gibson, Fred Janson, and the coordinators of several of the Round Robin groups got together and in 1967 devised a more workable alternative. The North American Fruit Explorers was established specifically to publish a quarterly journal, *Pomona,* to provide a more reliable means of group communication (Fishman 1986).

Although increasing membership has driven NAFEX to more formalized communication and a more formal organizational structure through the years, the idea of pursuing grants or membership growth goes against the grain of much NAFEX thinking. A basic theme throughout NAFEX's history has been the members' self-identification as amateurs and hobbyists. This, along with a strong sense of individualism, has been a major reason why they have maintained a decentralized organization. The board meets just once a year. The president has few duties other than to facilitate official functions, and only about 10 percent of the membership attends the annual meetings. The majority of the action takes place in local, informal groups, in fruit test groups, and between individuals.

This reluctance to build up a centralized organization and to allow it to be dominated by a few members is also allied with a strong desire to keep NAFEX from being dominated by professional fruit breeders or commercial fruit growers. A fair number of NAFEX members are professional breeders (e.g., breeders employed in universities and the USDA), and many more operate their own nurseries. They are valuable sources of information, advice, services, and plant material to other members. They are welcome in NAFEX despite their professional status because, as one informant put it, "many professionals are amateurs at heart," doing their work because they love it and not because of the money they make. But their professional and commercial interests are not allowed to dominate NAFEX.

An important influence on the strategy NAFEX develops is its interaction with the Seed Savers Exchange. Whealy at the SSE has been active in promoting the idea of assessing NAFEX's contributions to germplasm conservation by gathering a list of material held by NAFEX members. Whealy has been a member of the NAFEX committee to develop a database for fruit testing evaluations. He would like to see it used as the basis for monitoring germplasm conservation within NAFEX. His influence, along with that of like-thinking members of NAFEX, could have a big impact on any systematic conservation that NAFEX may implement.

ALHFAM

The Association of Living Historical Farms and Agricultural Museums (ALHFAM) is a group of institutions and individuals interested in interpreting agricultural history to the general public. There are roughly 850 members, located at about 60 historical sites. In the course of trying to create "living" exhibits to showcase past agricultural practices and artifacts, some of ALHFAM's members have found that heirloom varieties of vegetables, fruits, herbs, and ornamentals add a new dimension to their work. Heirloom varieties not only provide authenticity, especially when they can be traced in historical references, but they also give viewers of the exhibit a visual cue that agriculture has not always been the way it is today. Plants that look unusual sometimes prompt the public to make a comment or ask a question, which helps the exhibit staff find an opening for presenting their interpretation of history.

Living historical farms (LHFS) provide an opportunity for people to view a bit of the agricultural past and perhaps also provide a stimulus to think about the agricultural present. Historical farms provide this opportunity for education and reflection through the interpretation of artifacts such as buildings, clothing, furniture, farm implements, livestock breeds, and crop varieties.

At living historical farms, interpreting artifacts means showing them in use in historically appropriate ways and giving visitors the opportunity to discuss what they see. Such an interpretation could include a demonstration of the use of a spinning wheel, the use of a harness and plow with a team of horses, or the use of a scythe to harvest wheat. It can also include historically appropriate crop and garden varieties.

The use of plants to reconstruct historical settings varies considerably. At some sites, this is as simple as planting a kitchen garden with whatever species and varieties are readily available to remind visitors that gardens were ubiquitous in times past. At other sites, formal gardens may be planted to recreate the landscaping done by past inhabitants. At still other sites, the use of plants to interpret history goes to the length of researching which particular crop species and varieties were in use during the time period to be recreated, locating sources of those varieties, and planting them on the site. Although planting a kitchen garden with off-the-shelf vegetable seed is often sufficient to recreate an appropriate setting for the interpretation of other artifacts (e.g., buildings), the use of historically appropriate crop varieties makes a historical setting authentic and uses plants as artifacts themselves. Because some living historical farms are trying to achieve this depth of historical accuracy, they are interested and involved in germplasm conservation.

There is enough interest in the use of heirloom plant material within ALHFAM that the group has established a Committee on Seeds and Plant Materials. Several ALHFAM member institutions have researched and located plant material appropriate to the historical periods they are trying to reconstruct on their particular sites. Some institutions have begun maintaining material for themselves, while others rely on seed catalogs, especially catalogs from heirloom and "alternative" seed companies. There has even been a draft of a proposal circulating among some individual ALHFAM members to use living historical farms as sites for large-scale germplasm conservation.

At some historical sites, the use of heirloom varieties for interpretation is appropriate because a historical figure is associated with the site. The staff at Monticello, Virginia, for example, is trying to reestablish the orchards and gardens of Thomas Jefferson. Jefferson was a strong proponent of plant exploration and introduction and maintained large experimental gardens on which he kept extensive notes. The detailed notes and garden plans he kept are key to the modern reconstruction of the Monticello grounds.

For a plant variety to be used as a historical artifact, many ALHFAM members feel it is necessary to verify that it is of the same variety as was used on the site or in the area during the appropriate historical period. In other words, the plants should be heirlooms, descended from the same varieties mentioned in historical references. Facilitating historical research and locating modern sources of living material is one of the main functions of ALHFAM's Committee on Seeds and Plant Materials.

Verifying the historical authenticity of heirloom plant material is no mean feat. First, some reference to the varieties used on the site or in common use in the region has to be found. The next step is to try to locate a source of living material of the same variety. There are a few very old vegetable varieties still in common use such as Black-seeded Simpson lettuce, which can be found in any seed catalog or garden store. Others may be found at small heirloom, "alternative" seed companies. Another source sometimes used is the Seed Savers Exchange. In fact, some ALHFAM members offer seed in SSE's *Winter Yearbook*. Living historical farms also exchange material with each other. Yet another source occasionally turned to is the NPGS. Sometimes material is found under synonymous names, adding another difficulty in identifying appropriate material. Some material can be found under the appropriate name, but when grown out it fails to match its historical description. Finally, to add to the confusion, some old names have been appropriated for recently developed varieties.

A major impediment to locating authentic plant material is that few other germplasm collectors are so adamant about matching the proper name with the proper material as are ALHFAM members. At a minimum, an ALHFAM member would want to have the correct name and synonyms of a variety, the time period and region where it was used, its introduction date, and a morphological description for identification. Without accurate information, an heirloom is useless to them. Much of the material held by the NPGS, for example, is useless to historians because key information on where it was collected and what it is called ("passport" data) was kept haphazardly. Another problem with NPGS material is that some of it may be mislabeled. Heirloom material held by the SSE may at best be a little better. In SSE, however, members have sometimes encouraged each other to make up names for old varieties if they cannot remember the original name. Such advice must make ALHFAM members wince.[1]

Finding and verifying the historical accuracy of appropriate plant varieties can be a monumental task, the detective work that historians reputedly thrive on. If the task is carried through, a living historical farm can acquire an artifact that is as authentic as the old machinery, furniture, and buildings it maintains. Once a source of authentic material is located, however, the task may not be finished. Particularly for heirlooms that are not commercially available, a LHF may have to maintain the heirlooms it wishes to use. Those that do maintain their own heirlooms are at least implicitly engaged in germplasm conservation.

Some of the difficulties LHFs face in conserving heirloom material are ironically excellent lessons in the history of agriculture. Deer grazing in the corn patch has been a problem at the Homeplace – 1850 in Tennessee. Several consecutive years of drought have been hard on the wheat varieties used at the Oliver H. Kelley Farm in Minnesota. Problems with pathogens have made germplasm maintenance difficult at many sites. These problems illustrate to the public why these varieties have been largely abandoned, although the pathogens now decimating the heirlooms are not necessarily the same ones that were problematic in the last century. Although these problems yield instructive insights into some of the difficulties faced by agriculture in the past, they are distressing to the LHF manager who must save seeds to maintain them as artifacts. Some LHFs have lost valued heirlooms in this way, and others have been brought to the brink of losing their material. The irrigation systems, chemical sprays, and fences that could counter some of these problems are often not appropriate to historical interpretation. Moreover, they create selective pressures that may change the appearance of the next generation of plants. As a compromise, some LHF managers employ modern practices such as sprinkler systems and pesticides under cover of darkness. Many wish they had the personnel and the funds to establish separate germplasm maintenance plots away from the public viewing areas.

At present, then, ALHFAM as an organization does not conserve germplasm, although many of its members do. Some member LHFs have well-established plant programs, while others are just beginning to get involved. Given the opportunity, however, it would be appropriate to their mission to maintain heirloom crop varieties as artifacts. Agricultural museums could be an even more appropriate site for the maintenance of a large array of heirlooms than LHFs because their mission is not usually so site-and time-specific. Yet the most compelling reason for most LHFs to maintain heirloom germplasm is simply that the material would be unavailable for their use otherwise. Often obtained from obscure sources in very small quantities, these biological artifacts have to be multiplied and conserved for an LHF to display and interpret them. Members are interested in germplasm conservation both as a source of historically authentic plant material and as a historically appropriate practice.

NS/S

In southern Arizona there is a grass-roots conservation organization called Native Seeds/SEARCH (Southwestern Endangered Arid-Land Resource Clearing

House; or NS/S) which specializes in native and heirloom crop plants of the region. Plants such as tepary beans, chiltepines (a wild relative of chile peppers), devil's claw (used for food, oil, and fiber), and panic grass (the seed of which is ground for flour) are unique to the region's native agriculture. Devil's claw and panic grass may have been domesticated in the region. Varieties of these species are worth conserving not only because their long history in the area proves them to be adapted to local environments but also because they are part of the vanishing cultural heritage of the region. Like several of the other organizations described here, NS/S is attempting to ensure the survival of valuable plant species and crop varieties by getting them into hands that value them and that will use and save them. This includes Native Americans, gardeners, researchers, and other grass-roots germplasm conservation groups. Much of NS/S's work is ethnobotanical: finding out what plants Native Americans of the region use or once used and finding out how they were used, collecting them, and redistributing them to others who will continue their use. This includes wild, weedy, and domesticated plants. NS/S has also repatriated species and varieties that were once in use but have disappeared from particular Native American communities.

NS/S was founded by several agricultural development professionals and plant scientists. It got its start in 1978 when several Papago Indians tried to locate seeds of native crop varieties which their community had abandoned. Their request reached the Meals for Millions gardening and nutrition project, which in turn got in touch with a researcher at the University of Arizona who had been working with some of the plants they sought: Gary Nabhan. Nabhan, "delighted to find that his research might have practical benefit for the descendants of those Native Americans who had long nurtured this genetic diversity" (Nabhan and Dahl 1987:49), began working for Meals for Millions part-time. At about the same time, Mahina Drees joined Meals for Millions as a garden extension specialist, and together they started up the Southwestern Traditional Conservancy Garden. At first, this garden was located on Arizona Agricultural Experiment Station property, and later it was moved to the Tucson Botanical Gardens, where NS/S still has its offices and garden.

The goal of this Meals for Millions program was to help southwestern Native Americans "become more self-sufficient by providing them with crops that are highly adapted to the conditions in the region" (Jabs 1984:60). One of the basic premises of this agricultural development scheme was that "the health of agriculture in a region depends on these plants being grown where they are naturally adapted" (Nabhan quoted in Jabs 1984:61). By 1981 this Meals for Millions project had lengthened its name to the Southwestern Traditional Crop Conservancy Garden and Seed Bank and hosted an international workshop called

"Seed Banks Serving People." The sponsors of this workshop included Meals for Millions, the National Sharecroppers Fund, Pioneer Hi-Bred International, Inc., the Seed Savers Exchange, and Rodale Research Institute.

Because Meals for Millions could not fund long-term projects of this type indefinitely, in 1983 NS/S was incorporated as a nonprofit conservation, education, and research organization. Whereas the Meals for Millions project had a definite development focus, NS/S expanded its interests to include conservation of wild relatives of crops. NS/S has also expanded its research to folklore and nutrition connected with native plants. Most NS/S activities are carried out by a core staff and by volunteers. Their work is funded by the dues of 1500 NS/S members and money from the sales of seeds and publications. They also receive small grants, hold fund-raisers, and work on research contracts. Their current goals are to conserve native crops and related plants within the region and the culture of their origin, as well as to promote new uses for these plants.

Two major themes guide NS/S strategies for germplasm conservation, explicitly linking germplasm conservation with agricultural development. The first theme is the connection between biological diversity and culture, and the second is the connection between biological diversity and the environmental suitability of agriculture. Biological diversity and cultural diversity mutually reinforce each other. Different cultural groups select and use different plants in different ways, resulting in a diversity of plant species which are used and maintained within human society and in varietal and genetic diversity within species. At the same time, plants are a cultural resource that can help to maintain the cultural identity of a group through their symbolic value or through their use in culturally significant activities. In the Southwest, Native American groups use many unique plant species and varieties in their traditional agriculture and land use. Traditional practices are less environmentally disruptive than more recently introduced practices, taking advantage of plants that are better adapted to the region's climate and soils whereas introduced practices require more extensive environmental modification to succeed. Plant species and varieties that are traditionally used in the Southwest may therefore be a more suitable resource from which to create new cultivars for the region than introduced material. Additionally, traditional cultivation techniques and knowledge of plant-environment interactions may be a highly useful resource from which to build a more environmentally suitable agriculture for the region.

NS/S's conservation strategy has been designed in consonance with these themes of culture and environmental suitability. From collection to maintenance to getting the material to interested users, NS/S's conservation strategy is designed to maintain germplasm as both a cultural and an environmental resource.

The NS/S strategy for collecting germplasm targets plant material native to the arid Southwest, with a particular emphasis on plant species and varieties used in native cultural groups. NS/S acquires plant material in several ways. One way is through publicity and word-of-mouth. People hear of NS/S and offer plant material that they have and that might be of value to others. NS/S has also acquired material from NPGS collections. By identifying material in NPGS collections which was originally collected in the desert Southwest, NS/S has been able to reintroduce landraces of cotton and tobacco that have become extinct locally.

Another way NS/S acquires plant material is by making visits to Native American communities. Its representatives make roughly 30 such collection trips per year. Repeated visits ensure comprehensive collection of materials that are valued and used by a particular community and are adapted to their environmental setting. It also gives NS/S collectors an opportunity to ask what is growing well and to find out what plants the people in the community might be seeking themselves. They have visited some communities from 8 to 20 times over the last several years. NS/S has tried to identify areas that need collecting and is now in the process of mapping where collections have been made and where there are gaps.

NS/S targets a wide range of material on its collection trips. In contrast, USDA and IPGRI no longer fund multicrop collecting trips, instead sending out taxonomic specialists who receive grants from specific crop advisory committees. NS/S is sometimes ridiculed in professional circles for being interested in more than one crop. The tight commodity focus endemic to agricultural research makes it more natural for NS/S to work with botanical gardens and environmental groups, which have broader taxonomic interests.

An example of NS/S collection projects is a collaborative effort called the FLORUTIL Conservation Project (Nabhan 1988), involving Mexican, Navajo, and United States researchers from botanical institutions in the United States and Mexico. This project is primarily aimed at locating populations of plants which are traditionally gathered by native peoples but are now threatened by commercial overcollection as well as by desertification. The arid Southwest has been little visited by professional botanists, but the researchers involved with FLORUTIL have found that "native farmers and gatherers not only know the local distribution of rare species of agaves, sunflowers, and cacti, but that they sometimes protect, propagate, or watchguard these plants as well" (Nabhan 1988:1). This knowledge makes the participation of local people extremely valuable in conserving these resources on site. Previously, in a similar project assessing and comparing the floral diversity of Organ Pipe National Monument and a geographically similar area in Mexico, researchers found that diversity had been reduced since the native people had been forced to stop their traditional land use practices in the Organ Pipe National Monument area. In contrast, there was greater floral diversity on the Mexican site, where traditional practices persisted.

The germplasm conservation strategies employed by NS/S run the gamut from in situ maintenance of naturally occurring plant communities and traditional agricultural cropping systems to ex situ collections of material in botanical gardens and seed storage. NS/S participation in ecosystem conservation is primarily limited to identifying areas and plant species in need of protection, generally as part of joint research projects in which NS/S staff cooperate. Beyond this the organization does not have the resources to implement ecosystem conservation.

Some of NS/S's ex situ conservation consists of maintaining growing specimens of useful or endangered plants. This is necessary where in situ conservation might not ensure the protection of the plants, or where in situ collection does not make the material readily available to others who might want to propagate and use it and where the species in question is not easily stored as seed or is vegetatively propagated (e.g., some of the cacti and other desert succulents). This conservation strategy has been facilitated by NS/S's association with the Tucson Botanical Garden, which has provided the organization with office space and some land, and by its association with the Desert Botanical Garden in Phoenix, which recently established a display of plants used by Native Americans, in coordination with NS/S.

Ex situ seed storage and in situ agroecosystem maintenance form the largest part of NS/S's conservation strategy. In part, this is a legacy of its early work with cultivated plants and agricultural development, and in part it is a safeguard to ensure that rare material is conserved. NS/S has worked with frozen storage of seeds and pollen, ambient-temperature storage, and indigenous storage techniques. Clay jars of seeds, sealed with gum and tobacco, have been stored in caves and found to maintain good seed viability for 5 to 10 years. Encouraging native communities to use locally developed seed storage techniques is important because it is connected with their culture and also because it gives them local control and access to their resources. In addition, modern developments in seed storage have been examined and introduced to native communities where they might enhance local conservation. For example, NS/S has introduced the use of three-ply poly-foil bags and silica gel because storage in clay jars is not particularly amenable to maintaining a large number of varieties in one location. NS/S itself has begun to put seed in frozen storage in poly-foil bags at its Tucson offices and in the homes of NS/S staff. In addition, material distributed in bulk to native farmers or to gardeners through its seed catalog is stored in root-cellar conditions.

The point of conserving these materials is to maintain resources that are culturally and ecologically valuable. From the NS/S point of view, placing plant material back into its native context is the best conservation strategy of all: plant material that is in active and widespread use and is consciously conserved by many different people is not in danger of extinction. Genebanking is fine within its context. It has allowed the return of a cotton variety collected before 1910, which since had become extinct in native communities. But much of what is maintained in genebanks cannot be returned because it has been grown out poorly, or because its origin cannot be traced, or because native farmers do not usually have access to these collections. In contrast, in situ maintenance in active use gives local people access to their material and expands the range of material conserved. Such in situ conservation also serves to maintain local knowledge about the plant material and provides cultural continuity. It also maintains any adaptive interactions the material may have with other plants or organisms. In this respect, Nabhan has been quoted as stating that "seed banks are a back-up system" (Jabs 1984:61). From this pespective, distributing plant material is very important.

NS/S targets several groups of people to whom it distributes material, including researchers, gardeners, and native farmers. The first priority is to get material to those who will increase and save it. This generally means NS/S members and the native communities with which NS/S works. Material is given free of charge to Native Americans and to plant breeders. Some of the material is also offered for sale to the general public in an annual catalog, which helps support NS/S's activities.

In addition, NS/S staff conduct workshops, education programs, and public presentations to get the word out to local gardeners and farmers of non-native cultures that there is an indigenous pool of agricultural knowledge and plant material that may be particularly suited to their needs. To this end, NS/S gives tours of the Conservancy Garden in Tucson and gives slide shows and talks about native agriculture and gardening at local Native Plant Society meetings, garden clubs, and at meetings of scientific societies. In an effort to teach the general public how these plants might be used, NS/S staff also conduct workshops on the uses of native foods, for example, how to cook native meals. This public education serves several purposes. First, it makes the public aware of often overlooked resources and of their value. This is seen as an important step in generating public support for efforts to conserve such plant resources. Second, it may encourage some members of the general public to begin raising and using some of this rare plant material, thereby spreading conservation by getting more people actively involved.

RRC

The Rodale Research Center (RRC), a private nonprofit research organization, became involved in collecting and distributing plant germplasm in 1976. At that time Robert Rodale introduced a program to collect and catalog amaranth plant germplasm, to develop improved varieties, to provide information to farmers, and to encourage commercial production. RRC conducts research projects aimed at creating alternative cropping systems that are economically viable and environmentally sound. Amaranth, an underexploited plant with potential economic value (NAS 1975), was identified as a promising alternative crop in RRC's integrated farming system.

The need to maintain its own collection became especially apparent when, on a trip to recollect several amaranth samples in Mexico, the researchers found that the fields of amaranth had been replaced by urban sprawl. Genetic erosion in amaranth began with the arrival of the Spanish and has continued to the present. More recently, amaranth has been replaced with Green Revolution grains in Latin America and India. It is not regarded as a priority item by most agricultural research institutes.

As part of its amaranth project, the RRC maintained germplasm until 1990 and amassed over 1400 accessions (including nine elite lines developed by the Rodale Breeding Program) from 12 species of the genus *amaranthus*. Between 1982 and 1989 the center distributed samples of grain amaranth in response to requests from 607 organizations and individuals in 113 countries (Kaufman 1992).

Although the amaranth collection was meant to be a working collection for crop improvement, it became one of the world's most diverse collections. Because it was a working collection, it was first grouped into four broad categories according to use: grains, vegetables, ornamentals, and weeds, based on observations of the seed donors as well as observations made in the field at the research center. One of the major contributions of the Rodale researchers in germplasm characterization was the grouping of accessions based on common morphological features.

Because conservation was not a primary goal for Rodale and the center did not have the resources to implement a long-term storage program, its collection was transferred to the USDA Plant Introduction Center at Ames, Iowa, in the early 1990s. These materials were not always adequately stored, however. This experience illustrates the difficulties that even established private organizations like the RRC, in concert with government, have in adequately maintaining germplasm.

The RRC case also shows how germplasm conservation fits into crop improvement programs. The RRC is unusual in its selection of crop species and management practices, but it is very conventional in its conception of and strategy for germplasm conservation. Although in situ conservation and cropping systems might be ideal for controlling genetic drift in plant populations, moving samples to other locations facilitated plant breeders' access to the material. In addition, growing accessions in new environments revealed interesting characteristics of agronomic value. Indeed, most of the 1000 hectares of grain amaranth planted in the United States in 1989 was planted with the recently developed lines from the RRC program and the land grant universities. This case shows the plant breeders' imperative: conservation for use.

Conclusions

Striking similarities and differences can be seen upon examination of the germplasm conservation strategies of the organizations considered here. For each of the groups, germplasm conservation is only one of several activities. In some cases, although the group actually contributes to germplasm conservation, it may not be one of its stated goals. Individuals are almost always engaged in more than one use of germplasm simultaneously. A single home garden may be serving several of these uses. For instance, a garden may be a source of aesthetic pleasure, food, and income from seed production, all at the same time and with the same plants.

Among all the variations stemming from different meanings and uses of germplasm, three general types of germplasm conservation are recognizable: conservation of genetic material for plant breeders, conservation of crop varieties for gardeners and farmers, and conservation of natural populations for limited or indirect use. Although each type of germplasm is based in a different set of values and assumptions and each implies a somewhat different conservation strategy, the enormity of the task of conservation makes some interaction and possibly even coordination highly desirable. Although they hold different meanings of "germplasm," the commonality found in conservation provides a point from which a negotiation among conservation strategies may proceed.

Thus we can see that the NGOs complement rather than duplicate the activities of the NPGS, though both draw on various aspects of the larger national culture. Although no systematic comparison has been made for all crops, Whealy (interview) argues that relatively few accessions in the tomato collection of the SSE duplicate those of the NPGS. But neither the NGOs nor the NPGS have been able to develop fully effective systems of conservation. Furthermore, the very weak ties between the two systems makes the overall success rate lower than it otherwise might be. In particular, the very weak ties between the two diminishes support for the national program by limiting if not denigrating the work of the NGOs.

We would be remiss were we not to say something about the fragility of the NGOs. During our study, two – the Rodale Research Institute and the Grain Exchange – ceased conserving germplasm. Without greater public support and a wide range of funding sources it is likely that other organizations will succumb as well. Both NS/S and SSE have recognized this problem and have begun the difficult task of diversification in funding and leadership. NS/S has diversified its board to include Native Americans and a Mexican. It has reduced its dependence on grants to about half of its budget and has begun to create an endowment. It has also helped to establish the Traditional Native American Farmers Association, the first of its kind in North America. Nevertheless, there are many hurdles to overcome.

Finally, the disjuncture between food, agriculture, and environment tends to exacerbate the conservation problem. All Americans eat, but few are aware of how their food arrives at the table. Many Americans are concerned about the environment, but few are cognizant of the important – perhaps central – role that agriculture plays in modifying, maintaining, and recreating the so-called natural environment.

Yet, as we shall see, there are other ways to organize germplasm conservation, and though they are linked to cultural differences, that linkage is flexible, permitting multiple interpretations and multiple avenues of action. The following three chapters examine three cases that emphasize the complex but essential link between human cultural and crop genetic diversity.

PART THREE

Three Cultures, Three Agricultures

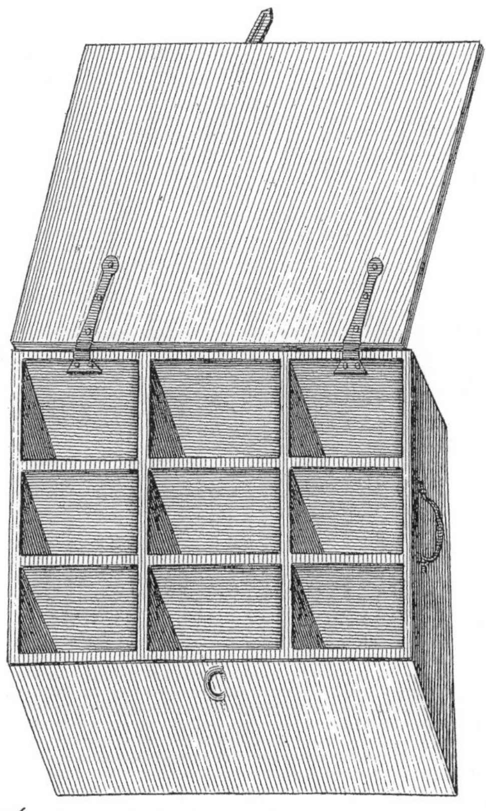

The Box with divisions for sowing different seeds in earth & cut moss from the southern Colonies and the West Indies.

In situ conservation generates its own
evaluation with every change of the season.

Garrison Wilkes

7

France: Remembrance of Things Past (and Future)

On the plains there will be no more peasants.
John Berger

Several years ago, a French newspaper reported on the Salon de l'Agriculture, the closest French equivalent to the U.S. state fair that exists: "A million people go each year to the [exhibition hall at the] Porte de Versailles; a record number that surpasses . . . that of the currently active farmers (988,000 in 1988)!" (Fottorino 1989:1). This simple statistic emphasizes the degree to which France has become an urban nation with a highly productive and science-based agriculture over the last 25 years. No longer does a trip through the countryside bring one past peasant farms and villages. More and more, the rural landscape resembles that of the United States.

The Salon de l'Agriculture features huge tractors, aisle after aisle of agricultural software dealers, and the glittering plexiglass cases in which seed companies display their wares. But the huge crowds come not for the tractors, the seeds, the prize hogs, or the irrigation equipment, but for the cheeses, wines, and *charcuterie* of France. Most of these people, like their American counterparts, have little if any firsthand experience of agriculture; for them, the Salon is as close as they will ever get.

As in the United States, however, the replacement of the variety of traditional agriculture with the uniformity of modern agriculture has led to the disappearance of many local landraces and weedy relatives of crop plants. In this chapter we examine the current state of plant genetic conservation in France. First, we describe briefly the historical role that France played in the dissemination of plants during the colonial period. Next, we look at the somewhat belated but novel attempt of the French government to deal with the problem of germplasm conservation. Finally, we suggest that the French model offers an important and challenging alternative to the American one, not only for the United States but for nations in the developing world as well. The key is that France came to recognize – albeit only recently – the inseparability of culture and agriculture. We begin with France's first foray into the plant sciences: the botanical gardens.

Le Jour de Gloire: The Rise of the Botanical Gardens

Much like the other major European powers of the seventeenth century, France saw in botanical gardens a source of potential wealth and power (Brockway 1979). Henri IV had a small garden with medicinal and ornamental plants kept for him on the Ile de la Cité, in the center of contemporary Paris. Even then, the potential value of control over plants was well known: the royal gardener frequently gave exotic flowers to the ladies of the court, but he kept the seeds for himself so as to safeguard his privileges (Bartélemy 1979). The French botanical gardens officially began in 1626 as a place to grow medicinal plants for the king. The Jardin Royal des Plantes Médicinales was located on the outskirts of seventeenth-century Paris. By 1641 it contained an astonishing 2360 species. In addition, the Jardin offered botany courses to the public free of charge and in French rather than the customary Latin. As such, the Jardin represented the avant-garde of science of its day.

By the eighteenth century, however, the gardens were seen not only as places where new crops for France itself might be developed but as staging areas for the transfer of crops between colonies. Guy Fagon, superintendent of the gardens from 1671 to 1718, encouraged exploration, importation, and "*acclimatization*" of all sorts of exotic plants, including coffee, tea, cocoa, and cinchona (for quinine). After 1726, a royal decree helped the process along: "His Majesty orders that the Captains and Masters of Navy Merchant Ships of NANTES who navigate in foreign Nations and the French colonies of America, will be required to bring on their return seeds and plants such as are found in the different places that they visit" (quoted in Guillard 1986). The impact of these plant transfers was considerable. For example, Brockway (1979:38) notes that "the French brought new crops to Algeria, principally grape vines and commercial citrus groves, but indigenous food crops suffered under an export-oriented economy." She also suggests that new crops were a major source of pre-1850 population increase around the world.

Under George Louis Leclerc, Comte de Buffon in the eighteenth century the gardens turned away from the search for drugs toward the scientific study of natural history. Though the revolution changed the name of the gardens to the Muséum Nationale d'Histoire Naturelle, it in no way slowed their growth. To the contrary, Paul (1985:57) notes that "from 1793 to 1840 the *Muséum* had been France's leading scientific institution, with a budget of 350,000 francs in 1825, while that of the faculty of sciences in Paris was 75,000." A single voyage by Captain Nicolas Baudin and four naturalists undertaken in 1796 during the throes of the French Revolution brought 100 crates of living trees and vegetables from Trinidad back to France (Bartélemy 1979). Moreover, through the Muséum, the French helped with the European project of the development of a scientific botany to replace the localized botanical nomenclature found both in Europe and in the colonies. In addition, the gardens helped to spread European plants (and animals) around the world, replacing those of local origin.

Nor did the French lag in the worldwide movement to establish agricultural experiment stations. As Busch and Sachs (1981) note, the French had established more than 84 experiment stations in the *metropole* and colonies by 1900 and more than 115 by 1930. Nevertheless, in 1881 only about 150,000 francs were being spent annually on research, compared to over 3 million francs in the United States (Paul 1985).

The French Revolution left most large estates broken into small plots and entrenched peasant agriculture. In 1815 almost everyone in France lived in the countryside, and most were agricultural producers. The average farm was 4.6 hectares, but in more populated areas farms of 3.0 hectares were more typical, and these often consisted of multiple parcels (Clout 1983). By 1880 the Jardin des Plantes was in such disarray that one of its buildings was literally falling down. And there were few opportunities for education for French peasants. Even as late as the 1950s, French agriculture remained peasant agriculture. As Clout (1983:153) put it, "the survival of a large landowning peasantry provided cultural continuity with times past and thereby ensured that change in the countryside was slow in the extreme." In this respect, the French scene was quite unlike that of Britain, where large estates were never broken up and the farm population had dwindled. And, of course, it differed from the large-scale, commercial agriculture of the United States.

What all this meant from the perspective of genetic conservation was that many of the old cultivars that had long disappeared in all but the most remote regions of the United States and Britain by the early part of the century still remained in relatively widespread cultivation in France in the 1950s. At the same time, however, the French lost their earlier concern for botany. Writing of the Muséum today, one author has stated that it "remains a little too attached to the brilliance of the period of Buffon, and not enough to the avant-garde work of its laboratories" (Bartélemy 1979:275). As one researcher remarked, France has lost its "*fibre naturaliste*" (interview).

The French approach to plant breeding differs somewhat from that in the United States. The French breeder, usually an employee of the Institut Nationale de la Recherche Agronomique (INRA), is expected to work on one or several crops, much like his or her contemporaries in the United States. Unlike his or her U.S. counterpart, however, the INRA scientist is expected to maintain a working collection of materials useful for breeding. The size, quality, and importance of these breeders' collections is arguable. Some breeders had and still maintain larger collections than do their American counterparts. And if our informants' perceptions are accurate, these collections have usually been transferred to new scientists as old ones retired. But the very informal character of the system makes it difficult if not impossible to assess it accurately. Some breeders insist that maintenance of collections was always a part of breeders' responsibilities; others argue that INRA concerned itself only with the production of new varieties (interviews). One breeder reported inheriting a collection of 1000 varieties from his predecessor of which only 500 are viable today because storage facilities were inadequate. Those that we saw varied from sophisticated cold storage and freezer facilities to drawers of antique cabinets in which unsealed containers of seeds were kept.

Until very recently, however, French scientists believed their collections were adequate. As a consequence, they perceived no need for a central facility such as that at Fort Collins to house large collections. The small collections used by breeders could be and were transferred on the breeder's retirement or transfer. This system differed considerably from that in the United States, where a lifetime of collecting, cataloging, and evaluating materials might be wiped out in short order upon the death or retirement of a scientist (e.g., Hymowitz 1984; Wilkes 1984).

Nevertheless, the French system for germplasm conservation was found lacking on several counts. First, it had no provision for long-term conservation of materials not currently of interest to breeders. Second, it lacked any method whereby a breeder or other interested party might easily learn about the materials stored in other collections in France. Third, it lacked any systematic program of exploration and collection. Of course, none of these were necessary as long as France had an agriculture that was largely peasant-based. What was not in collections was most likely growing in or around some peasant's field. In the

1960s, however, that situation ended when peasants' sons and daughters left the farms. The way of life that ensured the maintenance of local landraces and weedy relatives came to a close.

The response to this change was slow in coming and resulted in part from the international concern over the issue. More than one scientist remarked to us that they were surprised when they were suddenly bombarded by literature from the Anglo-Saxon world on the need for genetic conservation. Therefore, only very recently has genetic conservation been recognized as a national issue and even more recently has anything been done about it.

The Contemporary Challenge

Though in many ways the situation in France differs little from that in other industrialized nations, there are certain cultural and legal aspects that set it apart from its neighbors. Of particular import is that France adopted "scientific" agriculture only in the period after World War II. As a result, particularly in remote sites in the mountains, a considerable portion of the traditional landraces still remain in limited production. In addition, the French have three (and probably soon will have four) sets of separate and distinct laws pertaining to seeds. They also have considerable competition from foreign (largely American) seed companies seeking to furnish seeds for their domestic market. These have shaped the French seed industry and French agriculture in ways quite different from that of the United States. Let us begin by looking at the legal framework for seed sales.

The oldest currently enforceable law relating to seeds in France is a 1905 law on fraud and falsification, which was extended to cover the multiplication of seeds in 1949 and only in 1951 did it contain minimal rules for the commercialization of seeds. The law was modified in 1981, 1982, and 1984 as a result of changes in genetic knowledge and the exigencies of membership in the European Community (EC) (Lipschitz 1986). The law requires that for all seed commercialized in France, certain purity and germination standards be met, packages be sealed, labels meet certain specifications, and chemical treatments be noted on the package. Also, advertising must conform to certain established legal norms. The law is carried out by a system of regular sampling at sales points and points of importation or exportation. Some of the checks are accomplished on the spot while others require laboratory examination of the contents of the sampled packages. Seeds found wanting may be blocked from sale, and producers may be fined, jailed, or required to purchase newspaper advertisements explaining the situation to would-be buyers.

As late as the 1920s, the French courts insisted that seeds were not appropriable. Hermitte (1988) reports a 1921 case of a victim of World War I who had developed a new variety of carnation and had seen it stolen. He asked the courts for exclusive rights to it, but was refused. The seed industry, however, did succeed in establishing a Comité de Contrôle with a catalog of plants that were both new and improved in 1922. The Comité functioned as an unofficial agreement among producers until in 1932 a government catalog was established. It gave the creator some small protection – the public knew who the creator was – and it imposed various other obligations on sellers.

Today, all seeds sold in France must be certified and listed in the French catalog (or be listed in the European Economic Community's Common Catalog[1]). The Comité Technique Permanent de la Sélection des Plantes Cultivées (CTPS) is charged with the task of registering varieties in the catalog (Brossier 1986). The catalog was established as a way of protecting seed buyers from seed of poor agronomic quality. For a variety to be listed in the catalog, it must meet four independent standards: it must display distinctness, homogeneity, and stability – known as DHS – and it must perform better than other varieties in some clearly definable way. Garden crops are excluded from this last criterion. Distinctness means simply that the variety must be easily distinguishable from others on the market. Homogeneity means that the seeds must all yield plants that are homogeneous morphologically; it is tested by following the development of a small number of plants. Stability means that the variety must be homogeneous not only within generations but across them as well. Hybrid plants are given special treatment with respect to the DHS criteria; in particular, homogeneity is measured by reference to a test population of known homogeneity.

The system works by requiring each seed producer who wishes to sell seeds in France to request registration in the catalog. The requestor must make a case as to why the new variety is significantly better than others already registered. The seeds are then provided to the semiautonomous Groupe d'Etudes et de Côntrole des Variétés et des Semences (GEVES) for testing over a period of two years. (Temporary registrations are permitted for companies that have already successfully registered other varieties of the same species in the catalog.) Experts jointly designated by GEVES and CTPS report on DHS and yield tests. If the CTPS approves, the variety is listed in the catalog for a period of 10 years renewable indefinitely for periods of five years.

In addition, as part of a European cooperative program, some crops are not tested in France at all. For example, most forage crops are tested in England. Similarly, France provides the testing for German alfalfa. Moreover, there is a two-year delay between listing in the national catalog and listing in the Common Catalog so as to permit local testing; however, GEVES scientists note that sometimes the seeds do not arrive in time. Perhaps more serious is that varieties listed in the Common Catalog or other national catalogues have occasionally failed when imported into France.

Once a seed producer is granted a certificate and the seed is listed in the Common Catalog, enforcement procedures come into play. These are the province of the Service Officiel de Côntrole et de Certification (SOC) (Serpette 1986), whose authority is derived from the 1905 law on frauds. It is specifically charged with seeing that seeds conform to the description provided in the catalog. It does this by random inspections, taking samples for both field and laboratory analysis, although the sheer volume of varieties described in the catalog makes surveillance difficult (Serpette 1986). Moreover, the more narrowly that distinct varieties are defined, the more difficult it is to control what is in the market.

The third part of the legal framework surrounding seeds is of more recent vintage. Unlike the first two, which protect the cultivator, the third legal arrangement, the Droit d'Obtention Végétale (DOV), is designed to protect the seed companies. The DOV requires the demonstration of the same characteristics – distinctness, homogeneity, stability – as does registration in the catalog. In fact, the same organization, GEVES, is responsible for testing varieties for DOVs, but instead of presenting test results to CTPS, they are presented to the Comité de la Protection des Obtentions Végétales (CPOV). Created in 1971, CPOV provides certificates to breeders (usually as agents of seed companies) for new varieties that are either created or discovered.[2] Much like a PVP certificate in the United States, the Certificat d'Obtention Végétale (COV) grants its owner the sole right to produce and sell or to offer for sale the seed or other reproductive part or part used for vegetative multiplication for a period of 20 to 25 years. It should be emphasized that, unlike industrial patents, the COV limits its owner's rights to commercialization; anyone may use the registered variety for research purposes, including the development of another competing variety.

It is not necessarily the case that all varieties that receive COVs must also be listed in the catalog or vice versa. Thus, for example, a given landrace that is in the public domain may be registered in the catalog but would be ineligible for a COV. Similarly, a breeder's line designed as a parent for a hybrid might be granted a COV but would not normally be registered in the catalog because it would not be sold as seed.

Currently, there is much discussion in France and throughout the European Community over the court decisions in the United States that permit the granting of industrial patents for plants (chapter 10). Already the European Patent Office has accepted applications for a considerable number of plant patents. Although there is much disagreement over exactly what form a modification to the current system might take, it is clear that it will be decided at the level of the Community rather than simply in France.[3]

Without question, an industrial type patent for the products of biotechnology would create an unequal exchange between breeders (whose products would be available for further research without charge) and molecular biologists (whose products would be available only through some sort of license arrangement) (Hermitte 1989). This inequality would arguably tip the scales in favor of the chemical and pharmaceutical companies that wish to capture the seed industry. Not surprisingly, the established seed companies see little of value in the proposed industrial patents while the chemical companies see them as highly desirable (Joly 1989; Duesing 1989). Similarly, those connected with the Ministries of Agriculture and Research and Technology appear to favor a fairly limited form of patent for plants (and other life forms), while the French patent office favors following the American model.

All of what we have just described may give the erroneous impression that the current procedures for commercializing and claiming property rights for plants are relatively straightforward and simple. To disabuse the reader of that idea, a short detour is necessary into the complexities of the process. It should be remembered that all of this occurs after the seed in question has been designed through the conventional methods of plant breeding and has been submitted either to CTPS for inclusion in the catalog or to CPOV for a COV or both. Let us begin with the catalog, using feed corn as an example. In this case, the would-be holder of a registration would complete a three-page document that includes some 28 questions about the plant. These range from the date of the appearance of silks (very early to very late) to the color of the tip of the grain (white to yellow to orange to red) to the length of the peduncle (very short to very long). Each character is coded on a Likert-type scale with two to five categories.

Similarly, for a COV a seven-page form is necessary. It requests a list of already certified varieties that are similar to the one to be examined as well as a description of the specifically distinctive aspects of the new one. This is followed by 36 Likert-type questions similar to those requested for a listing in the catalog. These forms are then provided to GEVES along with 15 kilograms of seed. Samples of seed go to the two branches of GEVES, "seeds" and "varieties" (INRA 1982?). The "seeds" section engages in laboratory studies of seed purity, ploidy level, germination, sanitary condition, and moisture content. It uses tests that have been agreed upon by the Association Internationale d'Essais de Semences (AIES). The "varieties" section conducts field tests for DHS over a period of two to three years for both registration in the catalog and requests for COVS. In addition, for those varieties to be cataloged, tests of their "cultural value" and use are conducted.[4] In a given year as many as 500 requests for registration in the catalog and an additional 500 requests for COVS might be received. These tests require a network of over 75,000 parcels divided into more than 1000 experiments. In addition, because many of the tests require comparison with already existing varieties that are either in the catalog or possess a COV or both, GEVES maintains a collection of 16,000 reference samples on which it can draw. Maintaining this complex enterprise requires a staff of about 180 persons and a large budget, most of which comes from user fees.

Results of the various tests are then reviewed by committees of CTPS, CPOV, and SOC and actions deemed appropriate are taken. In the case of CPOV, the committee consists of a president who is a magistrate of the Court of Appeals of Paris and 10 members chosen for their competence in genetics, botany, and agronomy. They actually visit the fields and examine the written records of GEVES before making a decision as to whether to grant a registration.

Nor does jumping through the hoops needed for registration or certification end the process. Varieties do not simply stay distinct, homogeneous, and stable once they are made that way. This is particularly true for open-pollinated varieties for which variability must be maintained. In the past, farmers used various systems for collecting seed to plant the following year. For example, they might simply set aside a portion of the field for seed (in which case no improvement would occur). Alternatively, they might set aside a parcel on which they would select the plants that performed best (based on criteria that would vary by crop). Or they might choose the best plants over their entire field (interview). Seedsmen often did little more than this in their efforts to develop varieties to sell to farmers. The catalog, however, changed all that.

Landraces that were listed in the catalog were multiplied in situ for foundation seed. After that, multiplication for sale could take place anywhere. For perennials, seed could be harvested for up to three years. Then the process would begin again with multiplication from an initial sample in situ. This had the advantage of leaving the situation dynamic and evolutionary, but because it led to genetic drift, the practice is no longer permitted for open-pollinated crops. For closed-pollinated crops, it is still permitted, but "purification" on a plant-by-plant basis is practiced to remove nonconforming plants.

The newer and more cumbersome process involves beginning with a departure stock of fixed size. This stock is frozen and is used as needed until none is left. The stock that is withdrawn each year is multiplied again for a given number of times and then sold as seed. Thus, for example, if it is expected that a variety will remain economically viable for 20 years, one-twentieth of the stock would be removed from the departure stock each year and used for creation of the foundation seed. After 20 years none would remain and the variety would be withdrawn from the catalog. Though feasible in both principle and practice, the system is difficult to police; random tests can ensure conformity with the catalog specifications, but they cannot determine if the authorized methods of seed production were used.

The problems of maintaining varietal purity are apparent when one considers the case of the "18-day" radish. This landrace is currently listed in the catalog and is desired by some gardeners. By the late 1970s, however, it was noticed that most samples of it no longer conformed to the catalog description. The Ministry of Agriculture wanted to drop the name, but the seed producers protested on the grounds that open-pollinated varieties normally experience genetic drift and the name itself had commercial value among growers. In 1979 the variety was listed again in the catalog with acceptable limits of color and root length defined (Buret, Boulineau, and Brand 1986). Closed-pollinated varieties do not present the same technical problems but they, too, must be subjected to eternal vigilance if they are to be maintained as DHS varieties.

The other side of the legal coin of increasingly complex systems for bringing seeds to market is the changing French seed industry. France has the second largest seed industry in the world after the United States. Imports were only 50 percent of exports in 1960, but they were 115 percent of exports by 1980 (Grall and Levy 1985). A 1982–83 survey reported by Grall and Levy estimated that France had 765 seed companies of which 377 produced only and did not import. Of these, 112 had their own registered varieties and thus, presumably, some ability to do research. Twelve have research and development departments. Moreover, a few companies control very large shares of the French market for individual crops (SOLAGRAL 1988).

Many of these companies do not actually produce their own seed to sell to farmers because the processes of breeding and producing seed can be easily differentiated. The legal protection surrounding breeding makes it far more lucrative than seed multiplication. Thus many companies contract with farmers or farmer cooperatives to produce a given variety from foundation seed provided by the company under carefully specified conditions. Of late, farmers have been unhappy with the prices they receive for this work (Raymond 1985); in addition, labor to detassel maize has been lacking. This has been one factor in the movement of seed multiplication – but not the production of foundation seed – to certain Eastern European nations, notably Hungary (Grall and Levy 1985).

As in the United States, there has been a massive movement since the 1970s by large multinational petrochemical and pharmaceutical firms to purchase seed companies, following an earlier movement toward splitting the market into firms that were entirely French, those with American partners, and those under U.S. control. INRA, which had been a very important actor in the production of hybrids and finished varieties since its inception at the end of World War II, has seen its market share diminish considerably. INRA was forced to review its own policies in 1973 and to give exclusive licenses to one or more seed companies in return for a commission. More recently, INRA has formed consortia with seed companies around particular crops in an effort both to protect the French share of the world seed market (by giving seed companies a three-to-five-year advance on INRA lines) and to ensure that INRA itself has a role to play in the future. No single formula is used; rather, INRA policy is to treat each agreement separately depending on its likely impact on farmers and consumers. But INRA has had to be realistic about the growing role played by American and other foreign companies in French markets. These companies are now normally included in the consortia INRA establishes. In principle, the materials provided by INRA are to be used only to improve seeds for production in France; in practice, breeders admit that it is impossible to limit multinational companies' use of these seeds (interview). And the French have relatively few multinationals of their own.

The French seed industry is sharply divided into those firms producing hybrid seeds and those producing varieties. Despite the advent of COVs, hybrid seed production maintains a much higher return on investment because a farmer cannot save part of what he harvests this year to use next year as seed. Hence the small-grain sector – wheat, barley, oats, rye – remains largely the domain of relatively small, old, established French seed companies while maize, sugar beets, and certain vegetables are now the province of much larger companies. INRA has helped this process along by developing hybrid maize and vegetable lines and by establishing in 1983 its own foundation seed company, Agri-Obtentions, to sell its products to private companies for multiplication.

An example is hybrid maize, now an important crop in France for both feed grains and silage. Just after World War II American hybrids controlled the market, but the INRA hybrids released during the period 1957–60 recaptured it

(Cauderon 1983). In 1970 INRA held 78 percent of the maize market; however, by 1977, that share had been reduced to only 15.5 percent. Today, Pioneer holds fully half of the market. Pioneer multiplies its seed in Hungary and Romania for sale in the French, German, and American markets. This decline in maize seed production by French companies is of some concern because the high value added in maize seed has been used in the past by French seed companies to finance other less profitable research. In addition, France exports as much maize seed as does the United States, each having about one-third of the world market (Cauderon 1983); this is a world market share that the French would much like to retain. Pioneer's presence in the French market has become so important that one breeder remarked that one has to be careful not to overdo cooperation with it (interview).

In short, rapid internationalization of the French seed industry has forced a restructuring of public sector research, left France vulnerable in international markets, and increased competition in the industry. The legal changes described above played an important role in this restructuring. Moreover, the changes in both the law and industry structure have served to change breeding goals. Thus the design goals of the breeders have been changed by direct actions by new actors (e.g., multinational petrochemical companies that are now their employers), indirect actions of these and other actors to change the legal setting within which breeding is conducted, and the willingness of jurists to reinterpret legal traditions so as to conform to the needs and desires of various (new) actors.

All of the changes in the structure of the agricultural sector, the seed industry, and the nature of property rights accorded for plants described above have served to bring the problem of germplasm conservation to center stage in France as in other industrialized nations. The breeders' success has created concern for genetic conservation; the varieties they have developed have been so widely adopted that they have served to reduce variability in the fields to dangerously low levels. Thus the traditional varieties and landraces that marked French agriculture up until the 1950s have been replaced; many have already become extinct.

The concentration in the seed industry has added to the problem because large companies are ill-equipped to deal with the myriad microclimates and ecological niches found in every nation. Instead, they have bred varieties that can be used over wide geographic areas, further reducing the diversity in the field. Finally, the legal framework – the catalog and the DOV – have, through their insistence on distinctness, homogeneity, and stability, probably further served to narrow the genetic base in the field. The issue is not all that clear:

Marchenay (1987) has noted that the number of chestnut varieties has declined even though they are not listed in the catalog. Cauderon (1986), however, has argued that the catalog is concerned more with formalism than technique. He has suggested that the catalog be modified to permit sale to amateurs of unusual varieties of no economic value (Boucharenc 1988:4). In short, though no single factor can be held responsible for the changes, together they have served to define germplasm conservation as a problem.

Of particular note is the language used in the debate, which, by its very ambiguity, reflects the challenge of the new biotechnologies to conventional breeding. As in the United States, the French debate refers to genetic resources (*ressources génétiques*) – a term invented in 1967 (Rousseaux 1987) – which suggests on the one hand that what is of concern is not whole living organisms but individual genes and on the other hand that living nature is merely another natural resource. It is within this linguistic context that the French have attempted to develop a solution to the conservation problem.

The French Solution: Resources and Heritage

It is particularly ironic that France, a nation known since before Napoleon for its highly centralized bureaucracy and long before that for its propensity for grandiose (if often magnificent) projects, should undertake to conserve germplasm in a highly decentralized manner. But this is getting ahead of the story.

The first major government report on germplasm conservation in France was presented to the minister of agriculture in 1980 (Vissac and Cassini 1980). It noted three aspects of modern agriculture that led to the "impoverishment of genetic capital": specialization, marginalization, and pollution. It also noted that productivity gains to date were accomplished using only a very small genetic base. While breeding was concerned with the short term, genetic conservation was a medium- and long-term issue – but one with potential payoffs in the areas of better adapted plants, improved marginal areas, more complex agrarian systems, and biomass energy production. In addition, genetic materials were needed so that France could produce seeds for the Third World, a point to which we return below. It concluded: "It is a pity to observe that until now the *exceptional genetic capital of France has not been the object of a concerted policy of conservation that is equal to the problems posed*: a rapid change in the socioeconomic setting, [and] extremely lively international competition for the commercial exploitation of this capital" (Vissac and Cassini 1980:17). To remedy the problem, it proposed a national-level organization for documentation, research, low-temperature conservation, and regional coordination.

The report also contained an interesting appendix by the noted plant breeder Max Rives. He (1980:2) argued: "The importance of genetic resources has, at all times, been recognized by breeders: they have maintained for a very long time what they call modestly and without pretention *collections,* in order to be able to have permanently available sources of genetic variability that they know are necessary to their future work." Rives reveals his suspicions of molecular biology by arguing that, as it was developing, breeders were moving away from the concept of the gene and toward the notion of "genetic variability," a concept that emphasizes interaction among genes. Thus continuous progress in breeding requires not just one or two but perhaps hundreds of genes. Moreover, Rives warns against the notion of a genebank as a static conception; what is needed is not only storage but rational management. In concluding, Rives (1980:5) warns that "each nation must make its own effort, under the pain of falling under the dependence of the *varietal power* of other nations."

Despite this report, however, the story of germplasm conservation activities in France is very much linked to the efforts of one individual: André Cauderon. As a former maize and barley breeder, inspector general at INRA, and most recently permanent secretary of the Academy of Agriculture, Cauderon is considered by many to be the father of genetic conservation in France. Cauderon's evolving position on genetic conservation is a complex one and is far removed from that of most of his colleagues on this side of the Atlantic. In particular, Cauderon attempts both in theory and in practice to link an ecological concern for diversity with the practical and immediate concerns of plant breeders. Thus he argues that those species not immediately usable by human beings are not useless, for the diversity of the biosphere is essential for continued human survival.

Cauderon, however, rejects ecological romanticism (Cauderon 1980). He is aware that we can no longer count on nature or agriculture automatically to conserve genetic diversity. Positive intervention is necessary: "If we use adequate techniques, genetic standards and the search for higher yields do not make agriculture fragile. Productivity and regularity can be reconciled in diverse ways depending on the species; this orientation ought to be encouraged" (Cauderon 1980:1056). In addition, the conservation of diversity is the affair of the entire nation (if not the world) and not just of specialists.

Cauderon (1985) also suggested that two problems confront genebanks: diversity has no limits, and we lack the means to keep everything; and existing genebanks are far too little studied and used even though (or perhaps because) we know very little about genetic resources. In addition, he found French policy to be wanting with respect to the establishment of such facilities. To rectify the situation, he proposed that a center for genetic resources be founded that would coordinate research on cereals and include all concerned.

Of particular note is that Cauderon recognized the important philosophical and socioeconomic issues within the technical project of germplasm conservation as well as the political impasses that needed to be surmounted if the diverse interests concerned with the subject were to be united. For example, he argued that "to leave a place and sufficient autonomy to other living species while caring for the environment: in the land of Descartes, there is an uncommon interpretation of the 'mastery and possession of nature'" (Cauderon 1989). He also recognized that "it is not sufficient to conserve several living collections of seeds; it is also necessary to maintain alive the interest of Man for the treasure that diversity represents" (Cauderon 1984). To keep that interest alive, he has reached out to botanists, ecologists, and organic farmers (Cauderon 1981, 1982, 1986), among others.

His grand plan materialized in 1983 as the Bureau de Ressources Génétiques (BRG). He was named the first director. As it is currently constituted, the BRG occupies one floor of a small building on the grounds of the Muséum Nationale d'Histoire Naturelle in the center of Paris. The BRG contains no collections and no storage facilities. Only a small staff is present whose role is to coordinate the exploration, collection, inventorying, multiplication, and evaluation of the various collections scattered around the nation. The staff is paid by various research agencies. The Muséum provides the building free of charge, and the ministries provide operating costs. A Conseil d'Orientation helps set overall policy for the BRG. Its membership consists of representatives from the Ministries of Agriculture, Environment, Education, Foreign Affairs, Cooperation, and Research, as well as from industry and the major research organizations (i.e., INRA, CNRS, ORSTOM, Muséum, Veterinary Science). Its small size is underlined by the fact that it is not even technically an agency of the French government; instead, it is a private, nonprofit organization (Association 1901). This has the advantage of flexibility of operation but the disadvantage of being only loosely tied to any ministry and incurring the wrath of government accountants.

The BRG is part of a considerable and growing network of scientists and government officials both inside and outside France. It has organized several important colloquia on various groups of species. It has financed the creation and publication of catalogs of existing collections (e.g., Arbez et al. 1987) as well as several explorations. Representing France, it also collaborates with international organizations. In short, the BRG is both more and less than what Cauderon proposed: it is not limited to small grains but covers all living organisms. At the same time, it does not have the means to provide the needed long-term storage facilities for germplasm. Similarly, it can only encourage conservation activities; it still lacks the means to coordinate.

Given its newness, it is impossible to talk of a French germplasm system. To date, several initiatives have been taken. Some of them are described here to provide a more concrete view of the path taken. They may be divided for analytical purposes into two types of activities, which we call collection management and exploration.

With respect to management, the outlines of a structure are beginning to emerge. It appears that GEVES will be given the mandate to store items listed in the catalog. It already does this as part of its mandate as a support service for the CTPS and CPOV. In its expanded role, it would also be a designated genebank for these cultivars. INRA will probably house the collections for all the crops of major agronomic and horticultural importance. For example, work is already under way to develop a medium-term storage facility for small grains at the INRA station at Clermont-Ferrand. The botanical gardens and national and regional parks will be responsible for those items that are not of commercial interest.

The collection and evaluation programs under way are also worthy of mention because they involve some unique relations between INRA and the private sector. For example, the maize program has involved the triple financial support of the Ministries of Agriculture and Research and Technology, INRA, and 16 French seed companies grouped together under the name Pro-Maïs. Each of the three groups contributed about one million francs per year to a five-year project that ended in 1988. The seed companies' contribution was actually a contribution in kind; each company agreed to conduct a certain number of trials of collected materials and to report on them in an agreed-upon way. The magnitude of the project is indicated by the fact that 1200 landraces and varieties, about half from outside France, were studied. The entire group was studied in 29,972 plots. For each, their potential value for hybrid production (for either grain or silage) and "several million morphophysiological measures" were taken (Kaan and Boyat 1988). The results of the trials were then entered into an INRA computer and, based on statistical analysis, were regrouped into 36 pools for further work.[5] The initial results were distributed solely to the 16 participating companies and to INRA. Any other companies wishing to participate now must pay a Fr 200,000 entry fee.

Today, the collection is stored in a cold room (4°C and 35 percent humidity). Small samples are kept in a freezer (–20°C). In addition, the materials are believed by the breeders to meet the three requirements often cited in France for genetic resources: they are defined, maintained, and available. Future work will focus on evaluation and improvement of the pools formed as well as on methodological studies of the process of pool formation, but it will take place with a somewhat reduced budget because the companies and the ministries claim that they cannot maintain their earlier contributions. It is also likely that the companies are concerned not to go too close to the actual production of finished hybrid parents, where of course they are competitors. On the other hand, the Groupe d'Etudes des Lignées (GELI) exists within Pro-Maïs with the express purpose of using INRA lines to create its own hybrids. For this a royalty is paid to INRA. In addition, an expansion of the program to the EC level is being explored with particular interest in cooperation with Spain, Portugal, and Italy.

Two similar programs exist for wheat. Both differ in important ways from the maize program. The older program, in which only five private companies participate, the so-called Club de Cinq, has as its goal not so much germplasm conservation as improving resistance to disease and bread quality. Each company and INRA annually screen a wide array of materials. When interesting materials are found, all of the companies can work on them to develop new varieties. This program contains no common germplasm pool as there is for maize.

That pool is being constructed entirely by INRA researchers with no industry tie except through the advisory committee of the Office National Interprofessionel des Céréales (ONIC). The six INRA labs working on wheat have formed a network around a germplasm collection that is kept in cold storage at Clermont-Ferrand. A catalog has been established containing passport and basic agronomic data. A computerized database is being developed, and materials are being multiplied as necessary to prevent loss. The project began about 1985 with visits to all public and private wheat breeders in the country. (Private breeders were not asked to include material of "strategic importance.") This permitted the doubling of INRA's varietal diversity in a year (interviews). Future plans call for development of an evaluation network similar to that for maize. The approach to be followed is that the most exotic materials will be evaluated first at INRA labs and sent to the private sector only if found to be agronomically interesting. This practice is in tune with INRA's emerging policy of withdrawing from the production of finished varieties except when they form part of some more fundamental research project. All of the 25 to 30 small-grain breeding companies have agreed to participate in this program. Nevertheless, the ambivalence of the role of germplasm conservation in France is underscored by the fact that this program is currently being funded on a year-to-year basis.

The BRG also maintains links with scientific and voluntary organizations in other fields. For example, the BRG supported the printing and dissemination of a volume by Philippe Marchenay (1987) that provides a different rationale for collecting landraces and a guide to amateurs who might want to undertake such collecting activities. In Cauderon's preface to the volume, he states: "Many amateurs and associations wish to participate concretely in the safeguard of local varieties. It would be a mistake to underestimate their potential role: nothing solid can be done without the comprehension and the engagement of a large fraction of [public] opinion" (1987:12).

Marchenay, an ethnobotanist, takes the position that these now rare landraces should be kept because they represent a part of our heritage. Because they were created by our ancestors over centuries and they can survive only with our care, they represent a significant part of our past. Like an old painting or a historic monument, they provide us with an opportunity to appreciate the past. They can also provide certain of us with the pleasure of growing them, by following the traditional practices of our forefathers. These local cultivars were embedded in a sociocultural matrix that included systems for exchange of seeds, for communicating information about new varieties, and reasons for choosing one cultivar over another.

Moreover, this web of relations still exists in certain marginal and remote areas and can be understood in its own right. Marchenay suggests interviewing these persons to learn from them and to collect their seeds before they disappear. He warns against relying too much on old men because women often were responsible for gardening. He even provides forms to help the novice to collect and addresses where landraces can be sent for further analysis, multiplication, and storage. Finally, he provides a list of local organizations, of libraries with collections of documents, and of botanical gardens and parks with an interest in conservation activities.

Nor is Marchenay a single voice whispering in the wind. The national park at Ecrins has signed agreements with farmers to maintain in situ various local varieties of grain. A number of *ecomusées* have been formed around the country that serve to show the public how the use of agricultural space has changed over the centuries. For example, the Ecomusée du Pays de Rennes maintains 10 hectares with demonstrations of varieties, techniques, and lifestyles since the seventeenth century (Ecomusée 1988). The local agricultural college provides scientific advice on cultivation methods and varieties. In addition, INRA, several large seed companies, the local beekeepers cooperative, and the city of Rennes support the program.

The botanical gardens, too, have been enlisted in the growing network to support germplasm conservation. In addition to the Muséum, which furnishes office space for the BRG and collects landraces of grains, the botanical gardens of Nancy, Brest, and Porquerolles have medium-term storage for various groups of plants. All tend to focus on noncultivated species.

Nor should the rather large number of nongovernmental organizations involved in germplasm conservation activities be ignored. They include quasi-governmental organizations such as the Association Française pour la Conservation des Espèces Végétales, which includes among its members various research agencies; public-private groups such as the Association pour la Sauvegarde des Ressources Génétiques de Légumes et de Fleurs (INRA plus three seed companies); voluntary associations that do research and lobbying such as SOLAGRAL, a group concerned about the role of seeds in North-South relations; a half dozen Associations des Croqueurs de Pommes – literally, associations of apple crunchers;[6] and farm groups that are concerned about the loss of landraces (e.g., the biological agriculture movement).

Finally, the BRG maintains linkages with two unique government organizations whose mission is scientific and technical cooperation with the Third World. Both organizations have their roots in France's long colonial heritage. Unlike their U.S. counterparts, both ORSTOM (Institut Français de Recherche Scientifique pour le Développement en Coopération) and CIRAD (Centre de Coopération Internationale en Recherche Agronomique pour le Développement) maintain large staffs of scientists who are career civil servants. Moreover, both play a role in germplasm maintenance.

ORSTOM is France's basic research organization, with programs in the biological and medical sciences as well as soil science and the social sciences. Its Laboratoire de Ressources Génétiques et Amélioration des Plantes Tropicales maintains a 100m^3 cold storage facility (4°C) for seeds of various cereals, panicum, and gumbo, as well as in vitro collections of oil palm and coffee. In addition to attempting to develop improved methods of conservation, the laboratory is experimenting with the multiplication of plants from clones. This procedure is well established for oil palm and is under study for coconut. Another ORSTOM laboratory maintains a West African cereals collection (sorghum, millet, fonio).

CIRAD is France's applied research organization. It has a distinctly agricultural focus. It maintains considerable collections of germplasm for tropical commodities, including African maize varieties.

Finally, breeders and other scientists and administrators in France widely share the view that the central problem in North-South relations with respect to germplasm conservation is not one of control over genetic materials. Instead, it is the lack of adequate national agricultural research systems and trained plant breeders in these nations that can capitalize on the genetic diversity they have (Cauderon 1984; interviews). Thus time and again scientists recounted how duplicates of collections of materials made in developing nations were left with local authorities only to be neglected shortly thereafter. Despite the widespread acknowledgment of this problem, political and bureaucratic stumbling blocks make it unlikely that France will mount any major new effort to train breeders or provide financial support for Third World agricultural research systems.

In short, the BRG incorporates most of what was proposed by Cauderon. Indeed, the only major aspects lacking are the long-term storage facilities at the BRG, the laboratories for methodological research on conservation, and the technical advisory staff. To some extent, these shortcomings are compensated for by their availability elsewhere, but a sober assessment of the current French germplasm system would have to conclude that it is still a long way from achieving its initial objectives. Nevertheless, the French approach has merits over those taken by the United States and many other nations.

Why Conserve Germplasm: The French View

Any attempt to produce a French view of a topic as broad as the conservation of plant germplasm must undoubtedly caricature to some degree the amount of harmony that exists. Nevertheless, some major features do emerge that distinguish it from U.S. conservation programs. Let us consider them in turn.

FUTURE GENERATIONS

The most frequently given reason for genetic conservation in France is the same as that given in the United States: local landraces and wild relatives of crop species must be safeguarded so that future breeders will be able to continue to produce new and improved varieties. This orientation toward the future rightly puts the role of the breeder in the center, for it is the breeder who will use old varieties to fashion materials that can cope more effectively with new plant pests and can continue to increase yield over and above what it was in the past.

HERITAGE

The French position also draws heavily on landraces that are to be conserved for what they can tell us about the past. The national heritage (*la patrimoine nationale*) forms an essential part of the French approach. This position is widely supported by key figures associated with the BRG and by a surprisingly large number of bench scientists.

RESPONSIBILITY TO THE TROPICS

Finally, for more than a century the French have had a particular role with respect to the tropics. Some might consider this role to be sentimental – in both the English and French meanings of the word, that is, a longing for past glories and a moral commitment to the developing world. France is the only former colonial power to maintain several scientific organizations that have a strong tropical focus. The consensus across virtually the entire political spectrum on the need for such organizations can only be regarded enviously by American aid officials.

Yet France still clings to the remnants of its policy of making the colonial subjects into citizens by teaching them the virtues of French civilization. This is expressed most clearly in the two separate ministries that deal with the world outside France: the Ministry of Cooperation for the "French-speaking" nations and the Ministry of Foreign Affairs for the rest of the world. Moreover, there is a feeling of national duty toward the developing world and particularly toward the former French colonies that make up *francophonie*. This linkage is expressed not only in the form of foreign aid and technical assistance but in the form of what might be termed cultural aid, including regular international conferences on French language and culture. In the field of germplasm conservation, this sentimentality consists of French government initiatives to collect and store germplasm from the former colonies as well as to engage in plant breeding and associated activities by both ORSTOM and CIRAD.

Conclusions: Lessons to Be Learned

We can learn a great deal from the French approach to germplasm conservation because it is considerably different from those taken in the United States and other Western nations. In particular, the following points stand out:

1. It is near to the users. Rather than being concentrated in one or a few locations, as is the case in the United States, Germany, and Italy among other nations, the French have deliberately chosen to decentralize their system. Thus it is near to the largest group of potential users: plant breeders. Given that those physically present at the site of a genebank are far more likely to use it than others (Ashton 1988), this is a considerable advantage.

2. It aims to involve as many people as possible. By making the BRG an independent organization charged with creating a broad constituency, it has been possible to involve nontraditional clients in the process of germplasm conservation (e.g., ethnobotanists, amateurs, taxonomists, environmentalists). This contrasts sharply with the more limited American focus on the uses of germplasm for production agriculture.

3. It relates to both national and regional governments. In particular, regional governments have begun to realize that unusual varieties can play a role in local economic development and have been willing to invest in such programs.

4. It links national and international policy. Rather than having separate national and international policies for germplasm conservation, the French have opted to have one unified policy. This permits a level of coherence sometimes lacking elsewhere.

5. It acknowledges the inseparability of the technical and political facets of germplasm conservation. To French policy makers and administrators, the problem of germplasm conservation is at once a political and a technical issue. The way in which germplasm is conserved, who has access to it, and what is done with it are both technical and political questions. By acknowledging this link, the debate over germplasm is made more open and more comprehensible to the public at large.

But in France, too, the linkage between food, agriculture, and environment is becoming more fragile. Both the peasants and many local landraces have disappeared. In many long-farmed mountain areas, trees now cover ancient fields. And the supermarket and fast-food restaurant are rapidly changing the cultural landscape as well.

8

Brazil: Bringing Biotechnology In

> It doesn't matter who controls the international network of genebanks but who controls the "know how" for genetic manipulation.
>
> A Brazilian scientist

Contemporary Brazilian society reflects its colonial past and strong tradition of state intervention. In broad terms, there are two Brazils: that enjoyed by the elites and the middle class, and that known by the poor who make up the bulk of the population. For the elites and middle class, the separation between food, agriculture, and environment is nearly as complete as it is in the United States. In every large city, immense supermarkets supply a wide array of foods from tropical and temperate climates. Fast-food outlets can be found everywhere. Moreover, agriculture in southern Brazil is similar to that found in developed nations. Crops are produced on large, highly mechanized farms using the latest technologies. As a result, Brazil is now a pivotal actor in international markets for soybeans and concentrated orange juice.

In contrast, in the slums of the large cities and in many rural areas, especially in the northeast, food is often in short supply, crop yields are poor, and the environment is highly degraded. The relationship between food, agriculture, and the environment is all too clear to those who have been pushed to the side by modernization. Yet within Brazil are one-fourth of the plant species known on earth, giving it the largest portion of the world's biological diversity of any single nation (CIRF 1986). Nevertheless, like most nations, Brazil is dependent on plant germplasm imported from abroad to meet demands for food crops and to cope with the contradictions of modern agriculture.

Brazil requires the cooperation of many countries to gain access to their wild and improved plant germplasm. Yet to advance its position in the world market, Brazil has to compete for market shares and development of products with many of the same nations with whom it must cooperate. Hence, trapped in a cooperation-competition paradox, Brazil assumes an ambiguous position in the international debate over world plant germplasm.

The debate has been dominated by a "rich versus poor" scenario. Proponents of this view maintain that the North – developed nations, especially in the First World – is "gene-poor" while the South – Third World countries – is "gene-rich." It is implicitly assumed that a conspiracy by the nations of the North exists to exert control over the world's collected plant germplasm held in national and international genebanks. In this scenario Third World countries are expected to react as a united bloc to the threat to their botanical treasures. Conflicts from the scientific sphere will spill into the traditional economic conflicts between rich and poor nations. This position wrongly assumes the existence among the Third World countries of homogeneous economic goals, resource endowments, scientific capacity, and potential regional solidarity to counter the historical exploitation of their plant germplasm by developed countries from the North.

An alternative hypothesis can be constructed if we look into the various policy goals that differentiate the national interests of the nations involved, regardless of their designation as Third World or developed. Capitalists everywhere have become increasingly interested in the profit potential and market expansion provided by the biorevolution in agriculture. Biotechnology and genetic engineering also provide avenues of assault on problems in agricultural production that are advantageous to large corporations. In the West, the control of plant germplasm promises to take the place of the earlier battles over possession of land, with large-scale capitalists taking advantage of naive or weak governments and farmers in their quest for profits.

The question of control over these genetic resources is at the root of the conflict. This is why many multinational corporations fear that governments in some developing countries such as Brazil may have already started to implement policies to restrict or prevent access to their plant germplasm (e.g., Giacometti 1988).

The establishment of such restrictions violates the basic principle of the International Undertaking on Plant Genetic Resources (FAO 1987b), which defines plant genetic resources as the "common heritage of humankind" and establishes the importance of free exchange of plant germplasm among all nations.

Yet not all countries have an equal capacity to manipulate and transform the genetic information encoded in plants. Hence even the free exchange of plant germplasm does not guarantee equal advantages and benefits for all nations.

Free exchange of plant germplasm is important as a goal and should be pursued. Because of the far more developed scientific, technological, and industrial capacity of the advanced capitalist countries, however, free exchange alone will increase the benefits to industrial capitalists. It is not a coincidence that those capitalist nations (in both North and South) that have greater scientific capacity are supporting the ideology of the "common heritage of humankind" for plant germplasm (except for their improved, private germplasm) and at the same time promoting plant breeders' rights and plant patent laws worldwide.

To establish a more realistic understanding of the role of science in the political economy of agriculture as it relates to access to and conservation of germplasm, we need to describe the cooperation-competition paradox and to investigate the historical correlation between science and inequalities in germplasm exchange. We need to discuss plant germplasm exchange in Brazil since the colonial period and to analyze the Brazilian solution to the cooperation-competition paradox in that context. This is the focus of our analysis as we examine the place, role, and contradictions of Brazil's position in the "new world order" of postcolonial agriculture.

Plant Germplasm Exchange in Brazil, 1500–1995

Less than 10 years after the discovery of America, in 1500, Pedro Alvares Cabral discovered Brazil. When the Portuguese sent the first colonization mission in 1530, commanded by Martin Afonso de Souza, it brought soldiers, agricultural workers, tools, livestock, and seeds to Brazil (Editora Abril 1986:103).

Food crops alien to the native population were introduced later. They were imported to feed the colonizers and to assure the reproduction of the labor force, which consisted largely of African slaves. Commercial crops were introduced only to expand the market of the Portuguese empire abroad. Thus plant transfer from Europe to Brazil occurred only to guarantee a meal familiar to the first European immigrants, who came to the new land as part of the colonization project, and to feed the growing labor force necessary to spur the empire's economic competitiveness abroad.

Although plant transfer from Europe to Brazil started only in 1530, the first plant transfer from Brazil to Europe took place almost immediately after the discovery of the new land. Impressed by a colorful, floury pre-Columbian maize cultivated by the Brazilian indigenous population, the colonizers took it back to Portugal. Even today landraces of that maize can be found in Portugal (Giacometti 1988; personal communication).

The history of plant germplasm exchange in Brazil must be told within the context of events that unfolded under certain material conditions and were shaped by a specific scientific and political agenda. Basalla (1967) has argued that the spread of western European science may be divided into three phases. During Phase 1 science was marked by the gathering of systematic data using the categories of everyday experience. In Phase 2 scientists in the colonies or dependent nations looked to the center for scientific guidance. Finally, Phase 3 is marked by the emergence of an indigenous scientific community.

The case of Brazil follows closely Basalla's idealized pattern of scientific development. Phase 1 science took place in Brazil as soon as Europeans learned of the abundance of plant and animal species in the new land. Systematic plant hunting by European naturalists began in the seventeenth century. As Arnal (1987) identified in the case of Venezuela, they came, surveyed, took what they liked, and sent it back home without much regard to sharing any of their findings with the fledgling Brazilian institutions:

> Attempts to conduct scientific investigations in Brazil were made by foreign naturalists interested in gathering data on the fauna, flora, and natives, and on raw materials of possible commercial value. These "scientific explorations" were quite independent of Brazilian institutions. Their works were published abroad and placed in foreign libraries. Consequently, a knowledge of several aspects of Brazilian society could be obtained only through foreign literature. (Pastore 1978:268–69)

The rationale behind such a strategy was that a "plant monopoly" was the ultimate goal of European empires during the colonial period. First, there was an increasing demand for new food and medicinal crop plants. Second, the underdevelopment of the plant sciences limited their interests to crop end products (Souza Silva 1989). The list of nineteenth-century scientific expeditions in Brazil includes those of Auguste de Saint-Hilaire from France; Gregorii Ivanovich von Langsdorff from Russia; Wilhelm Ludwig von Eschwege, Johann Baptist von Spix, and Carlos Frederico von Martius from Germany and Austria; and Alfred Russel Wallace, Henry W. Bates, and Charles Darwin himself from England (Schwartzman 1978). Also famous was Abbott H. Thayer's expedition from the United States which brought Louis Agassiz (1867) and his collaborators.

Phase 1 science in Brazil extended well into the nineteenth century. Throughout the colonial period the establishment of scientific institutions was blocked by the Portuguese legacy. Isolated from the mainstream of European science, Portugal's own scientific establishment was very weak (Pastore 1978). This weakness led the Portuguese to restrict elementary education in Brazil. At the turn of this century, over 80 percent of the Brazilian population was still illiterate. Actually, science was utterly unimportant to the economic and religious elites of colonial Brazil. Only those institutions that were indispensable for the attainment of the economic objectives of the empire were allowed to be introduced (Schwartzman 1978). Thus the main educational establishments in colonial times were Jesuit colleges whose implicit mission was to diffuse the social, economic, political, and cultural values of the colonizers along with their religion.

Although Phase 1 science lasted until the turn of this century, the seeds of Phase 2 science were introduced by the Portuguese themselves soon after 1808. The rationale behind such an early move toward Phase 2 science was very simple. The colonial status of Brazil ended formally in 1808, when the country became a "united kingdom" with Portugal (Schwartzman 1978), although it achieved political independence only in 1822. The royal family and court fled to Rio de Janeiro to escape Napoleon's army in 1808. This "meant that for the first time . . . Brazil was granted the right to have a few institutions of higher learning and training" (Schwartzman 1978:546). The first agricultural research establishment, a botanical garden, was founded.

On 13 June 1808, the principe regente, later called D. João VI, authorized the creation of an Estação de Aclimatação – an agricultural station to permit the systematic introduction of plants brought from abroad. It became the Real Horto on 11 October 1808 and the Real Jardim Botanico do Rio de Janeiro on 11 May 1817 (Jobim 1988). Until Brazil's political independence in 1822, the botanical garden was used for the systematic introduction of foreign plants through methods that were not always legal. Avocado, breadfruit, and hog-plum were brought illegally by Luiz de Abreu Vieira e Silva, who escaped from the Ile de France (Mauritius). In 1812 Raphael Bottado de Almeida, who cooperated in the escape of Luiz de Abreu, sent the first tea seeds to Brazil (Jobim 1988).

Nor were plant transfers from Brazil to other nations always legal either. The most flagrant case was carried out under the auspices of the British Empire in 1873. At that time Henry Wickham, an English businessman, smuggled rubber seeds out of the country in response to a request from the director of the British

Royal Botanical Gardens at Kew (Galeano 1973; Brockway 1979).[1] Forty years later the British invaded the world market with Malayan rubber. Brazil, which had a virtual monopoly of the world market for natural rubber up to 1913, was buying more than half its rubber abroad a half century after the event. As a result, Henry Wickham is viewed in Brazil as a villain, whereas Francisco de Melo Palleta, who smuggled coffee seeds into Brazil in 1727, is seen as a hero.

The creation of other agricultural institutions in Brazil was influenced by the success of the European agricultural experiment station movement (Albuquerque, Ortega, and Reydon 1986a). "From 1859 to 1900, agricultural institutes were established in the states of Bahia, Pernambuco, Sergipe, and Rio Grande do Sul" (Pastore 1978:237). The most important of them was created by the central government in 1887 in response to economic problems emerging from coffee production. It became known first as the Imperial Estação Agronomia de Campinas and later would be called the Instituto Agronomico de Campinas (IAC). One hundred years later, IAC is still the strongest agricultural research institute in Brazil. Between 1887 and 1987, IAC scientists (now numbering 200) released 323 new cultivars for Brazilian agriculture (IAC 1987). Like that of all such institutions, it was marked by "a few scattered attempts at agricultural research . . . limited to replication of well-known European experiments" (Pastore 1978:237).

This statement supports Basalla's (1967) claim that Phase 2 science is dependent, strongly influenced by a scientific culture and tradition from abroad, especially from western European nations. Until the end of the nineteenth century, the members of the upper class in Brazil sent their offspring to be educated in Coimbra, Portugal. After Brazil's political independence from Portugal, however, interest began to shift to France. After the French Revolution, the Brazilian elites found the French drive to build a modern, efficient, and centralized state much more meaningful as a model of society than that of Britain. Later, Brazilian elites became so enchanted with August Comte's doctrine that "positivism was taken more seriously in Brazil than practically anywhere else. . . . [It] conferred on its exponents a sense of being modern, of being in favor of science and oriented towards the future, and it distinguished them from those obscurantists who thought that the ideal resided in the past" (Schwartzman 1978:548–49).

Thus positivism was pervasive in Brazilian social, economic, political, and cultural – and therefore scientific – life. Even a positivist church – Igreja Positivista – was established in Brazil. But most revealing is that the contemporary Brazilian flag still contains Comte's positivist slogan: *Ordem e Progresso* (order and progress).

European scientists also had a strong influence in shaping educational and research institutions in Brazil early in this century. Indeed, "the first Brazilian university was wholly European in tradition. France, Germany, Austria, and Italy provided the initial talent for both the University of São Paulo and the other Brazilian universities that came into being shortly afterward" (Pastore 1978:270).

Although until the early 1920s the state in Brazil was strongly influenced by the landlord oligarchy, late in the decade the hegemonic center of economic power shifted from agriculture to industry (Albuquerque, Ortega, and Reydon 1986a). This change brought about a shift in orientation, too. For instance, shortly after the Division of Genetics was created at the IAC in 1927, the institute sent an agronomist to be trained in genetics at Cornell University. Because the United States had already become a major center for the dissemination of Western science, Brazil decided to diversify its dependence: Brazilian scientists would also draw from the North American scientific culture and tradition.

Several themes characterize the conservation, use, and exchange of plant germplasm in Brazil from the time that the first agricultural research institutions were created in the last century through the early 1970s: lack of national policies; lack of a national system to organize and coordinate related activities; lack of a national program to integrate the existing efforts; and lack of an institution at the national level to assume most of the above tasks and to establish a bridge between Brazilian institutions and other countries. During the period marked by veneration of European science, such developments were virtually unthinkable.

Though Brazil is still strongly marked by elements of Phase 2 science, it is now struggling to enter Phase 3. When scientists from São Paulo founded the Sociedade Brasileira para o Progresso da Ciencia – SBPC (Brazilian Society for the Progress of Science) in 1949, the first seed of Phase 3 science was sown. But it was not until the military took over in the mid-1960s that definitive steps were taken toward Phase 3. These changes, however, were closely associated with the broader state hegemonic project whose ultimate promise was variously described as "the entry of Brazil into the developed world by the end of the century" (*Metas e Bases para a Acão do Governo, 1970–1971,* quoted in Aguiar 1986:73) and "putting Brazil, over the span of a generation, in the category of the developed nations" (Primeiro Plano Nacional de Desenvolvimento – IPND, 1972–74, quoted in Aguiar 1986:80).

For the restructuring of research, the United States rather than Europe provided the model for modernization. In 1972 the minister of agriculture, Luis Fernando Cirne Lima, nominated a special scientific committee to recommend reform of the national agricultural research system (Pastore and Alves 1984), which was coordinated by the Departamento Nacional de Pesquisa e Experimentação Agricola (DNPEA). The committee's report listed 10 shortcomings that were used by the government to justify its replacement. Obviously, those negative aspects would justify the restructuring and improvement of the system but not its complete replacement. Yet, it would have been inconvenient to the government to explain that there was a need to incorporate agriculture in the national economy to permit a more complete incorporation of the Brazilian economy into the world capitalist economy. Therefore, the real problems were not those listed by the committee (though they still needed to be solved). The establishment of a modernization mentality by the military demanded the establishment of totally new institutions and actors and a network of new relations so as to make possible the total integration of the country into the capitalist world. It is not merely a coincidence that most of the new institutions were named enterprise (*Empresa*).

Having already been affected by the two previous European institution-building waves – botanical gardens and agricultural experiment stations – Brazil was now strongly influenced by a U.S.-based third wave. Several Brazilian agricultural universities were created in the image of the U.S. land grant universities under the supervision of the U.S. Agency for International Development (AID). The agricultural research system under the coordination of DNPEA was replaced by a system first shaped for use by the international agricultural research centers (IARCs). Using the full weight of his international prestige, economist Edward Schuh, representing the Ford Foundation in Brazil in the late 1960s and early 1970s, played a critical role in the creation of what he called a "new model of research" – a commodity-based system of agricultural research resembling that of the IARCs (Albuquerque, Ortega, and Reydon 1986a).

The Present

The old agricultural research system was officially replaced in 1972 by the Empresa Brasileira de Pesquisa Agropecuaria (EMBRAPA) (the Brazilian Public Enterprise for Agricultural Research). EMBRAPA was the result of the most far-reaching and concerted effort the Brazilian government has undertaken to date toward Phase 3 science in the area of agricultural research. Thus EMBRAPA's first mission was the creation and support of state research systems and of commodity-oriented national research centers (Pastore and Alves 1984). The most striking characteristic of the new system under EMBRAPA's coordination was centralization (Albuquerque, Ortega, and Reydon 1986b). This was in conformity with the new planning and management style implemented by the military regime to give new direction to the process of accumulation in the Brazilian economy (Aguiar 1986). The philosophical and structural changes involving all major agriculture-related institutions had a profound impact on the conservation, use, and exchange of plant germplasm in the country.

Plant germplasm was defined as a priority. As a result, the National Center for Genetic Resources (CENARGEN) was created in 1974 for both plant and animal germplasm. This occurred at the same time that the International Board for Plant Genetic Resources was being created in Rome. Though plant germplasm has dominated its activities, since 1981 CENARGEN has been developing research projects aimed at characterization, evaluation, and preservation of "genetic groupings" of important local and naturalized domestic animals as well. Besides in situ conservation of animal germplasm, CENARGEN has also developed an animal genebank that includes conservation of semen and embryos through cryopreservation. The genetic groupings researched by CENARGEN include cattle, swine, goats, sheep, horses, and buffaloes (CENARGEN 1988). Researchers working with cattle genetic resources at CENARGEN's laboratory of experimental embryology have been successful in applying advanced techniques – micromanipulation of embryos – to obtain the birth of identical twins (*CENARGEN* Informa 1987).

In 1981, when EMBRAPA entered the burgeoning fields of biotechnology and genetic engineering, CENARGEN was selected as the site of the only laboratory in Brazil equipped to apply recombinant DNA techniques to crop plants. Thus, besides coordinating the National Program for Genetic Resources (PNRG), since 1986 CENARGEN has also coordinated the National Research Program in Biotechnology (PNPB). To provide CENARGEN with the capacity to perform its mission, the Support Unit for Research in Biotechnology (UAPB) was created with a three-module structure: molecular biology, plant cellular biology, and animal cellular biology. Thus though the acronym remains the same, today CENARGEN stands for the National Center for Genetic Resources and Biotechnology.

At CENARGEN, each director, coordinator, and researcher is also a curator for one or more crops or for a group of related crops (e.g., citrus). It holds the largest base collection for most crops in Brazil and coordinates a nationwide network of more than 70 active collections (called Active Banks of Germplasm – BAG) covering more than 70 different annual and perennial crops. The task of the BAG network is fourfold: multiplication, characterization, evaluation, and conservation of sexually and vegetatively reproduced crops. As the central institution in the system, CENARGEN controls the introduction, exploratory research, exploratory collecting, post-entry quarantine, clonal rejuvenation, conservation, quality control, germplasm-related documentation and information, and exchange of germplasm within Brazil and between Brazil and other countries (CENARGEN 1988).

CENARGEN is located in Brasilia and has 131 employees, of whom 53 are researchers (10 B.S., 30 M.S., and 23 Ph.D.). Thirty-nine employees provide research support (e.g., laboratory and field assistants, technicians). It has an experiment station and farm, greenhouses, and laboratories for research in plant and animal biotechnology and genetic engineering. It has storage capacity for 500,000 accessions. In the period 1987–88, 221 research projects were being carried out in the area of plant genetic resources by the Brazilian Cooperative System of Agricultural Research (SCPA); 89 were executed directly by CENARGEN and 132 more were conducted under its coordination (CENARGEN 1988). CENARGEN's researchers carry on research in 10 areas: introduction and exchange, exploration and collection, in situ conservation, ex situ conservation, characterization and evaluation, animal genetic resources, information and biometry, molecular biology, cellular biology, and biological control.

From 1975, when it became operational, until 1987, CENARGEN conducted more than 150 collection expeditions which produced almost 20,000 accessions (Lleras 1988). These invaluable plant genetic materials are now in the base collection at CENARGEN and in the active collections of other research institutions. The total number of accessions held at CENARGEN's base collection is about 40,000 (Giacometti 1989, telephone communication). CENARGEN now coordinates a network of 71 genebanks of which half contain germplasm for food crops and half are field collections of tropical crops (Giacometti and Goedert 1989).

CENARGEN also has seven genetic reserves and expects to create five more in the near future. It is working closely with the Nature Conservancy to document data on certain target species of particular importance including angico (*Anadenathera macrocarpa*), rosewood, mahogany, and Brazil nut (Lleras 1988).

There is no doubt that Brazil has increased its capacity to use, manipulate, and transform plant genetic resources since the creation of EMBRAPA. An example is the level of education among researchers before and now: "Under the old system, 10 percent of researchers held graduate degrees. The aim of the present program is that at least 80 percent of EMBRAPA's researchers will hold the masters' or doctoral degree" (Pastore and Alves 1984:127). In the fields of biotechnology and genetic engineering, the concentration of resources to upgrade Brazil's scientific capacity has been even more impressive. For the period

1985/86–1989/90, the National Research Council for Scientific and Technological Development (CNPq) granted about 8000 scholarships just for biotechnology-related M.S. and Ph.D. trainees. Furthermore, according to Furtado (1988), the Interministerial Council for Biotechnology (CIB) offered 3000 new scholarships to researchers wishing to obtain doctorates in biotechnology-related areas abroad. CIB wants to double the number of scientists in the field in three years: "US $50 million is being spent on graduate training in 1988 and CIB intends to double this in 1989 and 1990. This money is in addition to resources being devoted to training by each institution" (Furtado 1988:91). Moreover, CENARGEN has made an impressive move toward increasing the number of imported germplasm accessions in its base collection. In 1976, the annual number of germplasm accessions imported by CENARGEN was a little over 1500; by 1987 it had risen to almost 20,000 per year. Thus, perhaps Mooney (1985a) is mistaken when he implies that developed countries use more germplasm from the international network of genebanks because they exert a greater control over the network. Brazilian scientists, government officials, and research administrators assert that lack of capacity to use, manipulate, and transform plant germplasm explains why most developing countries request so few germplasm accessions from international genebanks. Brazil has established solid germplasm exchange relations with more than 30 countries (and it occasionally exchanges with other nations as well). CENARGEN gets over 50 percent of its imported germplasm from the international agricultural research centers. And among the nations of the world, the United States contributes the most (quantitatively) plant germplasm to Brazil.

National policies regarding conservation, use, and exchange of plant genetic resources in Brazil cannot be found in published materials, yet most of those interviewed agreed on what is or should be Brazil's position concerning the ethical and political issues addressed here. How could many researchers working in different research centers have such relatively harmonious positions if there is no explicit, official policy in Brazil to guide scientists and their institutions on the ethical and political aspects of Brazilian plant germplasm exchange at the national and international levels? The answer to this question is at the same time simple and complex. But both its complexity and simplicity are part of the strategy Brazil has adopted for resolving the cooperation-competition paradox.

Resolving the Cooperation-Competition Paradox

The International Undertaking on Plant Genetic Resources (FAO 1987b) mandated "the inclusion of elite and proprietary varieties of the North under the rubric of mankind's common heritage in plant genetic resources" (Kloppenburg and Kleinman 1988:295) (see chapter 10). Countries interested in influencing the direction of the international struggle had to position themselves with regard to the following outcomes of the FAO biennial conferences (Mooney 1989): the framework for an international system of genebanks funded and operated by national governments under FAO guidelines, the International Undertaking specifying voluntary guidelines for the conservation and exchange of germplasm, and the Commission on Plant Genetic Resources (CPGR), an intergovernmental body in which governments can discuss the issue. Because most capitalist countries in the North opposed the above measures, while most developing countries supported them, Mooney (1989:244) assessed this outcome as one of North against South: "All of those measures met stiff opposition from industrialized countries, especially the USA, Canada, the UK, and Australia. . . . Among Third World nations [supporting all three measures], the 'leaders' tended to be Mexico, Cuba, Colombia, Costa Rica, India, Pakistan, the Philippines, Senegal, and the Congo, though support was broad and enthusiastic, particularly from Latin America." But in Latin America, the largest and most influential country, Brazil, did not align itself with the bloc from the South. The only plausible explanation for Brazil's position is that, given its ambiguous position in the current changing world division of labor, it regards prompt adherence to the Undertaking as strategically inconvenient to its ambitions as a candidate to the select club of the Western capitalist powers. And it is very likely that Brazil will support the commodification of plant germplasm in the near future, by means of plant breeders' rights or plant patenting laws.

Indeed, under the leadership of International Plant Breeders (IPB), a giant seed producer controlled by the British/Dutch company Royal Dutch Shell, with the support of the Brazilian Seed Association (ABRASEM), Brazil's largest hybrid seed company, Sementes Agroceres, and a few scientists, PBR legislation was almost approved in Brazil in 1977. Paschoal (1986) provides an excellent account of how the project resulted from collusion between IPB, ABRASEM, and Agroceres. But strong opposition on several fronts formed by some scientists at the Agronomic Institute of Campinas (IAC), São Paulo's House of Representatives, and the Association dos Engenheiros Agronomos de São Paulo defeated the project. EMBRAPA was involved in the episode by a request from the Ministry of Agriculture.

Yet Brazilian scientists, administrators, policy makers, government officials, and businessmen are aware of the strategic importance of some economic activities and of some raw materials. All indications are that they have been unofficially developing a concerted effort to establish implicit policies that may help them to deal with some highly sensitive issues. These include the need to have access to the plant germplasm of those countries with which Brazil will be competing and the need to manifest its position on the FAO International Undertaking.

The rationale behind the implicit policy approach adopted in Brazil is very simple. Most scientists, administrators, and government officials interviewed view Brazil as having one foot in the Third World and another in the First World. Most of them believe that Brazil is going through a critical transition period from developing to developed nation, which sometimes brings about both economic and political dilemmas. In a given context it is more convenient for Brazil to assume a posture closer to that of other developing nations. In other contexts such a posture would be very inconvenient. But even when a Third World posture is not convenient, it does not mean that Brazil should explicitly assume a First World position. That might be tantamount to political suicide (and vice versa). Thus Brazil's ambiguous position in the international debate reflects its alliance with developed countries in opposition to the FAO's definition of what plant genetic resources must be considered "common heritage" and must therefore be made available to all nations.

It is in this context that one can understand why Brazil has become a member of the FAO Commission on Plant Genetic Resources, like most Third World countries, but has not adhered to the FAO Undertaking. But it did not declare itself against the Undertaking, as did Australia, Canada, Japan, Switzerland, and the United States (FAO 1986a:3). Brazil's representative at the 1983 biannual conference at FAO asked for time for Brazil to present its official position. To date Brazil has not adhered to the International Undertaking, although it is a member of the commission. It has been rumored for some time that Brazil would reveal its official position as accepting with restrictions. This would put Brazil on the side of all Third World countries that have adhered to the Undertaking; but its restricted acceptance would align it with countries such as Belgium, Denmark, Egypt, Israel, the Netherlands, New Zealand, and the United Kingdom (FAO 1986a:3). This position would avoid the inconveniences of simple opposition.

An excellent example of how Brazil is acting in the policy area is that explicitly it does not put any legal or official restriction on exchange of plant germplasm. Implicitly, however, it has restricted access to the germplasm of its most important native crops: pineapple, cashew, cocoa, Brazil nut, rubber, *guarana* (*Paullinia cupana*), and *caiuae* (*Elaeis oleifera*). The exchange of germplasm of these crops is restricted (Giacometti 1988), though no Brazilian government document states such a restriction. All those interviewed were unanimous in asserting that Brazil does not prevent access to the germplasm of those crops; it restricts their exchange only to increase its bargaining power vis-à-vis other countries that prevent access to plant germplasm of interest to Brazil. For instance, India has approved a law that prohibits the exchange of its black pepper germplasm, but it wants cashew germplasm from Brazil. Because of its strong interest in obtaining black pepper germplasm, Brazil has unofficially conditioned India's access to Brazil's cashew germplasm to India's permitting Brazil to have access to its black pepper germplasm. Brazilians cite the cases of restricted access to Indian pepper and Ethiopian coffee germplasm as excellent examples of "inconvenient," "misguided," and "distorted" policies. They defend Brazil's implicit policy of restricting access to the germplasm of its most strategic native crops not as an indicator of Brazil's desire to prevent access to it, but as a way of increasing Brazil's power to negotiate access to germplasm of interest with countries that try to prevent access to it. As one Brazilian scientist put it, "Implicit policies give more political flexibility to a country to act on a case-by-case basis." The germplasm of "restricted exchange" is always available through bilateral agreements, which is how Brazil is resolving the cooperation-competition paradox.

Conclusion

In a world strongly marked by international competition, global interdependence in plant germplasm affects all countries and regions (though it affects some more than others). Countries need access to the wild or improved plant germplasm of countries with which they may have to compete. Because every nation is dependent on others, but also has to compete with many of those others, developed as well as developing countries are trapped in the cooperation-competition paradox.

Because modern agriculture's most critical wild plant germplasm is located in developing countries, the greatest scientific capacity to manipulate and transform them is located in developed countries, and developed countries have historically exploited the plant treasures of tropical countries, critics of the existing inequalities in the international pattern of plant germplasm exchange have defined the struggle over plant germplasm as one of North against South. An examination of the case of Brazil, however, challenges this "them" against "us" approach.

Though Brazil's botanical treasures have historically been exploited by Western capitalist nations, the restructuring of the country's model of capital accumulation in the 1960s and 1970s led Brazil to improve its agricultural scientific capacity as a necessary condition for its incorporation into the world capitalist economy and for increasing its international competitiveness. This is what led the country to improve its capacity to use, manipulate, and transform plant germplasm. As a result, Brazil is now among the few developing countries with a chance to challenge developed countries in the fields of biotechnology and genetic engineering applied to germplasm-related problems in the tropics.

As industrial capitalists and the capitalist state in Brazil have become aware of the potential for profits represented by the new possibilities opened up by molecular biology, they have become aware that the biorevolution means also a gene revolution. But because they are also aware of the need to cooperate with those with whom they have to compete, Brazilian scientists, policy makers, industrialists, businessmen, and government officials have been applying implicit policies to bypass the cooperation-competition paradox. That explains the unofficial policy of restricted exchange of plant germplasm. Furthermore, by becoming a member of the FAO Commission on Plant Genetic Resources without adhering to the FAO Undertaking (or adopting it with restrictions), Brazil challenges the erroneous assumption that the struggle is one of North against South.

One needs only to read how in the 1970s IPB, ABRASEM, and Agroceres almost got breeders' rights legislation and patent laws approved to understand the struggle as shaped by capitalist interests in both developed and developing countries. And sooner or later Brazil will approve breeders' rights or plant patent laws. CENARGEN has already created the Brazilian System for the Registry of Cultivars (SBRC). It has been accused of opening the door for the introduction of plant breeders' rights legislation because it might permit a public or private research institution to make a legal appeal. CENARGEN's SBRC has been strongly opposed by some EMBRAPA researchers, especially those at the National Soybean Research Center (CNPSO).

At the same time, grass-roots initiatives for the conservation of plant germplasm have grown considerably. In the last five years more than 100 NGOs have been established, mostly in the Amazonia region. These organizations have yet to establish a national voice, and they are not linked to any governmental body. Moreover, rumors abound that interests from the North have created some fake NGOs to screen medicinal plants, translate indigenous knowledge, and gain access to germplasm illegally. Whether true or not, the rumors demonstrate the growing awareness among Brazilians of the strategic importance of germplasm for Brazil's economic future.

In short, no single country or region is self-sufficient in plant germplasm for modern agriculture's major crops, and this global interdependence and the existing global economic competition have trapped all nations into the cooperation-competition paradox. The spread of Western science to the rest of world has also set the stage for the inequalities that have characterized the struggle over plant germplasm from colonial times to the present. The increasing scientific gap between developed and developing countries is as problematic or more so than access to plant germplasm that most of them do not have the means to use, manipulate, and transform. As one Brazilian scientist remarked to us, "Developing countries should be more concerned with improving their capacity to transform plant germplasm than with discussing who will coordinate an enormous amount of germplasm which most of them are not prepared to use." Even the free exchange of all types of plant germplasm (including private sources) would not equally benefit all nations because only developed and a handful of developing countries have the capacity to take advantage of such a policy. Interdependence in plant germplasm and economic competitiveness are affecting both developed and developing countries. Finally, Brazil's solution to the cooperation-competition paradox and its ambiguous position in the FAO debate reveal the pitfalls of the North-South approach adopted by some.

A geographical entity – the North – does not exploit another geographical entity – the South. Only people exploit people. The attempt to privatize plant varieties in Brazil is illuminating because it shows that, when it is convenient, capitalists from developed and developing countries are able to unite their forces, with the support of some in the scientific community and in the state apparatus, to advance their projects, even though they may have to compete with each other in the future. The neglect of this dialectical interplay over what both nature and society will be has been the major conceptual weakness of most of those involved in the debate – even those defending the cause of the South.

9

Chile: A Latecomer Enters the Debate

> When I started to study agronomy, I realized that plant germplasm is the basis for feeding the coming generations; so that conserving genetic material seems an imperative. That is why I am in this business.
>
> A Chilean scientist

The primary reason why Chile's plant resources should lay claim to world attention is the enormous amount of natural variation among Chile's crop plants. Modifying the findings of Vavilov, Jack G. Hawkes identified four main regions of genetic origin and ten of genetic diversity for crop plants. A region of diversity is where the family to which a crop plant is related shows the most genetic variation (Hoyt 1988). Chile is the backbone of one region of genetic diversity.

Because Chile is 4300 kilometers long and has unique topographical features, it is endowed with great climatic and soil variations. The Andes, reaching altitudes over 6000 meters, run the entire length of the country. The mountains divide Chile from east to west, and the distance divides her from north to south into five distinct climatic and agricultural zones. The Central Valley, the heart of Chilean agriculture, separates the Andes from the coastal mountains. This is the most productive area, though its average width is only 160 kilometers. The five agricultural regions, the Great North, the Lesser North, Central Chile, the Lakes Region, and the Southern Region, stretch from north to south. Each of these regions has its distinct precipitation, temperature variation, and humidity. No wonder that modern crop plants should show all their variation here.

The climate has also led to wonderful variations among plants. The ability of these plants to develop is also affected by varied land use practices. Over 30 percent of the land is used as range and forest land. Slightly over half of the tillable land is cropped or planted; the rest remains in natural pasture. Principal crops are wheat, maize, beans, potatoes, rapeseed, and various fruits and vegetables. Traditional agriculturalists have used Chile's climate to develop an impressive variety of tomatoes (*Lycopersicon*) and beans (*Phaseoulus*). Weedy relatives in the wild and forest plants on the 30 percent of Chile covered by undisturbed flora add yet another dimension to its genetic diversity.

The development of seeds and genebanks in Chile must be seen in the context of Chile's colonial rule by the Spanish, its development as an independent bourgeois state, and subsequent civil war traumas (Moraga-Rojel 1989). The Spanish were apt to give the care and protection of the native peoples on royally granted lands to the holders of the *encomieda* (trust). This meant that not only were the lands given away, but the indigenous population was given away as well. The trust and later the *merced,* or grant, created a system of large estates and sharecropping by the end of the seventeenth century (Moraga-Rojel 1989). Through the concentration of ownership, the budding Chilean aristocracy was able to produce for the imperial Spanish and international markets in gold and tallow. The essentially commercial motivation of the Chilean conquest and the distinctly export orientation of the early colonial economy contradict its characterization as feudal, autarkic, or merely subsistence in character. "Capitalist" would be a misnomer as well, however, because the Hispanic juridical order was coupled with various religious and cultural institutions to make Chile "a miserable poor backwater of the Spanish Empire" (Loveman 1988:74). Private enterprise and private profit depended greatly on coercive labor, the royal protection of monopolies, price fixing, and dedicated markets. There was no significant wage labor force until the end of the nineteenth century.

Against this sociopolitical backdrop the issue of Europeanization arose. Just as the Spanish had imported the social system that was being imposed in Chile, they also wished to import the food system to which they were accustomed. The plant importations and trade were to differ from those of the United States, but the issues of socioeconomic development and plant development were inexorably linked.

During the nineteenth century, Chile continued its development as an exporter of metals, developing more copper and silver mines. Commercial agriculture turned to this small home market, and the potential for export to Australia and California under a monopoly position in the Pacific grain markets (Jobet

1955; Sepúlveda 1959). This advantageous position gave the landowners the income and political muscle to increase their holdings and their control of agrarian labor (Góngora 1960; Kay 1977).

The *inquilino,* or tenant system, in the Central Valley was the heart of oligarchic power. In contrast, tenure from Talca to Concepcion was based on smallholdings, independent millers and farms employing free wage labor. These entrepreneurial farming interests were coupled with certain mining interests against the holders of seignorial land. Despite this opposition, the manorial interests were able to withstand the banning of slavery and *encomiendas* and convert free settlers into explicit labor service tenancies (Zeitlin 1984). This allowed the landowners to increase greatly their surplus for export.

The 1850s saw the advent of capitalist agriculture as labor was drawn into the coastal region's mining and railroad development. By 1860 rails linked the Central Valley to ports, and Chile began to export wheat to Europe. The estate owners responded by increasing the work and decreasing certain benefits (e.g., meals, grazing rights, and housing) of those *inquilinos* already bound through debt to the land. Moreover, they were reluctant to mechanize. As one observer has put it, "while the wheat flowing to Europe from the plains of Kansas was mechanically cut and threshed, that from the Chilean Central Valley and coastal range was produced in the traditional way" (Wright 1982:5).

The Agricultural Exposition held in Chile in 1869 spurred large landowners to modernize. They founded the Sociedad Nacional de Agricultura (SNA) at that time with a mere 147 members. During its first 20 years the SNA received a state subsidy and acted as a quasi-official ministry of agriculture. Only in 1922 did it have its first membership drive with a goal of 2000 members, and that was done in response to growing class conflict and the rise of Marxist parties (Wright 1982).

Zeitlin maintains that the system was unique in that it combined features of the estate from feudal relations and the firm of capitalist development. This is evidenced by the integrated production of agricultural commodities under typically capitalist conditions. Indeed, the flour mills of the hacienda system were technologically among the best in the world at that time (Zeitlin 1984). These mills were owned by the mining interests in most cases and indicate the extent to which the earlier battle of capital and land had been won by the landowners.

The large estates that were either bought or owned by the mining magnates were to enable Chile's entry into the age of mechanized agriculture. In the Central Valley, the mine-associated lands were the first to acquire machinery, irrigate, and introduce new crops based on yield for the market (Encina 1955;

Zeitlin 1984). Hence landed property became agrarian capital and landlord and capitalist became one. This fusion posed internal questions about which way Chile should forge into the future. While the landed oligarchy viewed land as a source of status in itself as well as a source of power and domination, the emergent modern agriculturalists viewed the control of capital as primary and viewed land as a means of control, not as an end in itself. The latter's belief in controlled technological progress as a determinant of efficiency and thereby of market control was to play a decisive part in the future role of science in Chilean agriculture.

Genetic Erosion and Cultural Erosion

Seed saving and even genebanks are not new to traditional agriculture in Latin America. Plucknett et al. (1987) report that the Kayapó Indian women of the Amazon maintain germplasm collections on hillsides. Representative samples of the genetic diversity of major food crops are constantly maintained and protected by elderly women under the direction of the female chief.

These practices in the Amazon are similar to the indigenous germplasm development of Chilean farmers, especially small farmers who normally save seed for more than one growing season and who select and store many varieties of seed to hedge against the variable climatic conditions. In traditional Chilean cultures seed durability, its ability to winter over, has long been a traditional descriptor in seed selection.

Unfortunately, in the last several decades Chile has lost much of its crop diversity. Until the mid-nineteenth century there existed ample genetic diversity in all sexually reproduced plants in Chile. Economic pressures as well as changes in consumption habits, however, have led to widespread genetic erosion and the replacement of traditional cereals and vegetables by new, improved, commercial varieties.

The process of genetic erosion is difficult to measure. One problem is that there are only estimates of the number of plant species that were historically present in Chile. In the Vavilov "regions of diversity," however, there is evidence that important crops for food production are suffering from genetic erosion. Two such plants are the grain species *Bromus mango* and *Chenopodium quinoa*. These plants were staples in the pre-Columbian native diet and, like many native cultures, are facing extinction today. *B. mango* can no longer be found in Ibero-America, and *C. quinoa* has been eliminated from the Central Chilean breadbasket. These species were put into their precarious position by their replacement with high-yielding wheat cultivars brought by the Europeans. This is the unfortunate consequence of bending the land to imported tastes instead of developing the necessary institutions and research program that would allow for the development of new higher-yielding varieties of these indigenous crops.

The causes of genetic erosion in Chile are manifold. In the commercial agricultural lands, the accelerated replacement of traditional cultivars for those used in the advanced countries has damaged the integrity of the local landraces. The replacement of local varieties by improved ones has been faster in this zone, from 28° to 42° south latitude, than anywhere else in Chile. Many small farmers in this commercial zone have changed their production habits to conform to the dictates of the market, replacing their ancestral varieties with commercial ones. Native species have also been lost through the socioeconomic transformations that Chile has experienced in the last several decades (agrarian reform and increasing rural-urban migration). The loss of subsistence farmers and the integration of Chile into the world economy has created the structural context for the elimination of small-scale agriculture, allowing for capital-intensive farming, including the purchase of technologies and seed based on yield alone. The destruction of "marginal" virgin forest and agricultural areas to clear the way for livestock production, with its attendant economic advantages in the market, has promoted the elimination of many habitats for native plants (Moraga-Rojel 1990). Finally, the aggressive national plant breeding programs, spurred by multinational seed and pharmaceutical companies, have accelerated genetic erosion by encouraging the adoption of uniform cultivars.

Native crops from the indigenous civilization are not the only crop plants suffering from genetic erosion. The most recent collections of maize between 18° and 42° south latitude also indicate severe genetic erosion (Paratori 1988, personal communication). Out of 472 classified accessions only 15 percent correspond to typical landraces, 27 percent to weedy relatives, and the remaining 58 percent to hybrid lines, all of which present a strong introgression of genes coming from other landraces or from commercially available varieties. Genetic erosion has been more severe in areas where commercial hybrids were cultivated nearby. This is one of the reasons why traditional varieties of maize are likely to be found only in the extreme northern and southern edges of cultivated areas. A similar situation has been noted for potato germplasm. The Isla Grande of Chiloé, considered by Vavilov as a center of genetic diversity, is no longer a viable source of genetic material for potatoes. Researchers now are forced to travel farther south into the few remaining subsistence economies to find potato germplasm that might be used to prevent another catastrophic blight.

There are other causes of genetic erosion as well. One is the presence of plant diseases in the soil. Some imported plants are killed by these diseases; sometimes the imported plants are hosts that develop more virulent strains of the disease. In certain cases, indigenous plant life is killed by imported pests and diseases. The development and extension of these plant pests and diseases have wiped out several sources of genetic materials for breeders. Potato blight, lentil rust, barley yellow fluting, wheat rust, beans yellow, and common mosaic virus have all claimed broad areas of cultivation as victims. There is no way of estimating how many species have been weakened to extinction through their import. Nor is much known about genetic erosion of other species of Andean cultivars. Vegetables, subtropical fruits, and forage plants compete with one another for a place in the biosphere, and some are excluded in the long run. At any rate, commercial crops of these species are based exclusively on improved genetic materials.

National efforts to collect and to conserve traditional cultivars have relied heavily on the researchers' own interests and curiosity. Although these efforts to collect and conserve genetic diversity have expanded during the last two decades, they are still ill-funded and poorly organized. Chilean agricultural officials had overlooked the issue for a long time. Only recently have agricultural scientists, aware of the implications of the international debate over genetic resources at the FAO, as well as that over patent protection, succeeded in convincing political authorities of the importance of conserving and using agricultural diversity.

Structure and Organization of Genetic Conservation

Despite the immense genetic variety of its countryside, Chile does not have a national system devoted to collecting, evaluating, and documenting plant genetic resources. There are certain institutions that may perform these tasks, but a comprehensive program does not exist. This situation contrasts with the priority and scope of programs in the United States and many Western European countries. It can be explained by the differences of Chile's role and experiences in the world market for foodstuffs.

Many Chilean agronomists are aware of the policies needed to launch a national germplasm effort. The most notable efforts in focusing and coordinating germplasm conservation are those carried out by the Faculty of Agricultural Sciences at the Universidad Austral de Chile (UACH). A group of researchers from the UACH Crop Production Institute has carried out the collection, conservation, evaluation, and documentation of potato germplasm for over two decades. Since its inception in 1964, the National Agricultural Research Institute (INIA) has extended this effort to collect native maize germplasm. Finally, the

National Forestry Corporation, CONAF (Corporación Nacional Forestal), maintains a collection of native and improved forest plant seeds at its experiment station in the nearby city of Chillan. Among those private organizations involved in plant collection are the National Farmers' Association and the Baer Seeds Enterprise (centered on lupine). The remaining agricultural institutions and professionals engage in tangentially related activities. Preservation occurs in the context of specific plant breeding programs.

The first national symposium on germplasm conservation was held in 1984. It was spurred by the international debate over plant genetic resources within the FAO during the early 1980s. This symposium was held under the auspices of the Universidad Austral de Chile and was sponsored by the IBPGR. Representatives of the Chilean plant breeding and genetic engineering community attended. They recommended the formation of a national commission to be charged with developing and coordinating activities connected with phytogenetic resources nationwide. The dreams of agricultural scientists concerned with germplasm conservation crystallized into plans, organizational procedures, and goals when a formal commission was created in late 1986.

One of the first concerns of the INIA-led commission was the creation of a national genebank. To this end INIA received a mission from the Japanese International Cooperation Agency (JICA) in 1989. An agreement was signed establishing a joint project in 1990; it was renewed in 1993 for three more years. The project's objective is to advise INIA officials on the formation of the Plant Genetic Resources Project (PGRP). Soon after signing the agreement, Japanese representatives visited potential sites for genebanks. In 1990, they agreed that the Vicuña locality, in the Lesser North region of Chile, was the most suitable place for the base collection. The project, funded by the Japanese with some Chilean support, intends to create one long-term storage genebank at Vicuña, three active banks in Santiago, Chillán, and Temuco, an in vitro collection, and one quarantine bank in Santiago. As part of the project, scientists are involved in a professional exchange and training program with both Japan and the United States. In addition, in 1989 a biotechnology laboratory was established at INIA's La Platina Experiment Station. The laboratory focuses on the use of germplasm through tissue culture and molecular biology. Support for equipment acquisition and exchange of genetic material are also parts of the project (INIA 1987; Cubillos 1988, personal communication; Cubillos and Suzuki 1991).

Genetic Erosion and Germplasm Collection

The Plant Genetic Resources Project is making use of the existing plant collections in Chile. The main pre-PGRP collections were of economically valuable material for breeding. In addition to the potato collection at the UACH, the maize collection maintained by the Maize Program at INIA (La Platina Experiment Station), and the collection of forest seeds at the National Forestry Corporation in Chillán, there is a collection of forage plants for arid zones at the University of Chile (Las Cardas Experiment Station). Other minor collections of germplasm are maintained in several public and private institutions and are linked to specific plant breeding programs. Wheat, maize, and bean collections are maintained at the National Farmers' Association's Graneros Experiment Station, near Santiago.

In the last several years, collection activities have been carried out by taking into consideration international standards such as site, kind of material collected, topographic observations, type of soil, cultivation conditions, and the environment in which the plant grows. This international standardization of plant germplasm descriptors is ironic. The steady loss of genetic diversity in Chile can be understood in the historical context of the development of capitalism within which science and technology were used to exploit the Chilean genetic heritage. Now, to become independent of that historic development, Chilean scientists are cooperating with scientists from their historically rival nations to save their genetic heritage for future generations.

A poignant example of forced dependence is the Dutch seed company ZPC (Cooperative Association for Improved Seeds). In the early 1980s this company bought a Chilean seed company as a Latin American base of operations. Because the ZPC varieties are sold in "technological packages," only the larger farmers are able to afford such technology (Hobbelink 1987b). This situation displaced native species of potatoes. At the same time, small farmers – local *campesinos* – were at a technological disadvantage compared with large farmers.

Crossing colonial plants with European varieties has increased the wealth and diet of European agriculture for years. For instance, resistance to virus X of potato was located in a landrace of the Chilote people of Chiloé Island; this resistance has been incorporated into cultivars in Scotland and in the United States. Another landrace of the Chilote people was used as a parent in a potato line that has been used widely in the former German Democratic Republic and Soviet Union for resistance to virus Y and the potato wart (Plucknett et al. 1987).

Despite the recognized value of germplasm conservation, the germplasm collections in Chilean institutions are subject to a variety of problems, mostly affecting the ex situ germplasm storage system. To agricultural scientists at INIA and the university, ex situ conservation has acute limitations, especially in cases of vegetatively propagated crops and recalcitrant seeds (Contreras 1987; Paratori 1988, personal communications). In fact, conservation of germplasm far from its source of origin has become the major procedure for preserving crop genetic resources in Chile. The discontinuance of in situ preservation on the basis of costs incurred has stripped the dynamics of natural development in plants from the scientific community. But the ex situ position is highly conducive to the breeders' needs because the material in genebanks often has already been classified and evaluated to some extent. Yet conserving genetic material ex situ is an expensive enterprise requiring trained personnel, stable policies, and minimal personnel turnover. Many of the most crucial positions within these institutions are held by individuals who are nearing retirement and for whom there are no readily identifiable replacements.

Chilean agricultural scientists agree that in situ germplasm conservation is a very controversial issue because of the philosophical and political questions involved. They are concerned with the dilemma surrounding traditional cultivars as modern crops take over. It is paradoxical that some national policy makers are eagerly promoting the preservation of traditional cultivars while there are other political priorities and pressures to adopt improved crop varieties and the associated fertilizers and other inputs. In situ conservation is usually seen by development advocates as an attempt to keep indigenous groups of small farmers in a state of agricultural suspended animation. Despite these criticisms and apprehensions, many of the agricultural scientists are convinced that in situ conservation remedies many of the mistakes brought about by the Green Revolution, such as excluding small farmers from the benefits of high-yielding technology and subjecting them to the subsequent genetic erosion of their primitive cultivars. In situ conservation serves multiple functions for agricultural communities and keeps a sample of any germplasm collected in the hands of interested farmers. In situ conservation can also serve as a focal point for the community and act as a teaching tool as well so that farming practices and associated cultivars can be passed on to the next generation of farmers. Moreover, this also serves to increase the awareness of regional food producers of the importance of their agricultural achievements.

The San Juan de la Costa potato project, carried out by the UACH Crop Production Institute's staff and funded by the International Development Research Centre (IDRC) of Canada, serves as an excellent example of in situ germplasm conservation integrated with the development needs of the Huilliche Indians, and other small farmers as well, in the coastal zone of the Lakes Region (Contreras 1987; personal communication 1988). This socioecological approach, involving farmers directly in the process of germplasm preservation, has been copied in the northern part of the country with indigenous plants, particularly quinoa, maize, and regional fruits.

In addition, the Centro de Education y Tecnologia (CET) has been engaged in an in situ conservation program on the Chiloe archipelago. Through this program local farmers help in maintaining exotic potato cultivars in an area reputed to be one of the centers of diversity of the crop (Montecinos and Altieri 1991).

Finally, Chilean agricultural scientists and administrators interviewed concluded that the in situ approach is complementary to the already established ex situ germplasm banks and to the planned national genebank. The latter is seen as an important step toward diminishing Chile's dependence on foreign sources of germplasm.

A recent dispute over germplasm requested from the Netherlands is a vivid example of the need for Chilean plant breeders to establish their independence. In 1986, Chile had difficulties in obtaining disease-resistant strains of a bean variety from Holland. This germplasm was finally released after intensive negotiations by INIA, emphasizing the need for Third World nations to increase their bargaining power.

The bargaining chips that are most valuable now are the negotiations surrounding the Biodiversity Convention and the International Undertaking. These guidelines for genetic exchange and conservation will be the cornerstones of a new and hopefully more equitable international arrangement. Chile, although endorsing the International Undertaking without reservation, as yet has neither defined policies for dealing with its plant genetic resources nor devised a clear program to protect them. Currently, resources are not subject to any restrictions except for those tree crops that the National Forestry Corporation (CONAF) has labeled "national protected patrimony." Among those crop species are *Araucaria araucana* and *Libocedrus chilensis*.

Access to Chilean genetic material is available upon request to CONAF. CONAF regulates both the import and export of germplasm, together with the Agricultural and Livestock Service (SAG). A duplicate of the collected species must be left in those Chilean institutions that cooperated with the collection team. Importation of germplasm is subject to phytosanitary and quarantine regulations provided by SAG. The genetic material entering the country must be submitted to International Phytosanitary Certificate inspection to prevent the entrance of pests and diseases. There has been an active exchange of plant genetic materials with numerous national and international institutions extending over the years. In this respect, INIA has followed the long-standing tradition of exchanging its maize and wheat germplasm collections with the National Seed Storage Laboratory at Fort Collins, CIMMYT (Mexico), and other research centers in Argentina, the United States, France, Germany, Spain, and India. The potato germplasm collection has been sent as seed to England, the Soviet Union, the International Potato Center (CIP) in Peru, and other places. INIA's breeding programs keep up an active exchange of germplasm with the corresponding programs of the international agricultural research centers. Most of these reciprocal exchanges do not fulfill the conceptual definition of a network because these nurseries and trial activities are defined at the international center level with very little interaction with the participant institutions. In some instances this generates conflict because scientists working in the national centers feel credit due them for their work is attributed to, or appropriated by, an international agricultural research center, leaving them in obscurity (Venezian 1987:96).

Chilean universities and private research entities are not included in the international exchange network. Their collaboration with international centers is casual and of a personal nature. For the most part, these contacts are not on a programmed basis and are often mediated by INIA. Most of these exchanges are simply responses to requests by foreign institutions. For instance, the Universidad Austral de Chile's potato breeding program uses CIP materials, which it gets through INIA's channels. It is a bit ironic that the university system, which has the most complete collection of germplasm in the country, has to go through INIA to obtain internationally traded materials and is excluded from the international network for plant genetic material focused around the IBPGR or the IARCs.

Today Chilean authorities are concerned about the advance of modern technology, especially that of genetic engineering, and its potential to increase agricultural productivity. Mendel's genetic gravy train is racing forward, and no one wants to be left behind. In the international race to develop biotechnology, Chilean officials as well as scientists are especially anxious. As Plucknett et al. (1983:163) put it, "They feel that the industrial revolution passed them by; [so] they see biotechnology as a way to catch up and they are doing everything they can to be part of the genetic revolution."

One step in this direction was the transformation of one of the two traditional universities in Talca city (Central-South) into a genetic engineering university (Jorge Urzúa, personal communication, 1988). Scientists at Talca University are now working on plant micropropagation and establishing linkages with the private sector, especially with Chile's profitable fruit production sector.

Prospects for the Future

Chilean agriculture developed as an extensive system, taking advantage of the abundant natural resources. Water, a temperate climate, expanses of fertile soil, and genetic diversity allowed production to grow without the intensive measures needed in European agriculture. Nevertheless, even such abundance cannot go beyond certain physical, technical, economic, and ecological limits. The country has been approaching these limits for some time.

The current problem facing Chilean agriculture involves the need for an urgent and direct relationship with research and technological advances. This relationship, however, as recognized by plant scientists, requires political support, capital investment, publicity, and education of qualified personnel so that the latest knowledge is relevant to and sufficiently used by the eventual beneficiaries. To paraphrase a scientist, unfortunately for the people who are making political decisions, it is very difficult for them to understand that research is not something in which we are going to have results next year. With constant and determined support, research results are obtained, risks are reduced and usefulness for the country and for its constituents is optimized (personal communication, 1988). For plant scientists, one aspect of importance is the availability and adequate use of plant germplasm. By having a strong genetic base, either in situ, ex situ, or both, the country could develop a sustainable agricultural research program for the coming decades.

There has been an active exchange of materials with international institutions and other countries, but almost all agricultural scientists are concerned that free access to genetic resources is becoming increasingly difficult. The number of countries and institutions refusing requests has grown steadily over the last few years.

Chilean agricultural scientists have also shown concern about the current international debate over the rights of multinational seed companies and traditional farmers. They are aware of the legal and ethical conflicts that have powerful repercussions for individuals and organizations dealing with the conservation of plant genetic resources. Much of the concern expressed by plant scientists is fueled by the belief that international seed companies use traditional farmers' germplasm as the foundation for the development of their improved varieties. These are then sold at a profit to large Chilean farmers. But Chilean scientists and administrators did not support the suggestion that such gene traffic be shut down. They see closing the borders and forbidding foreign collecting expeditions as a suicide measure. They maintain the fairness principle of free access to germplasm despite the vulnerability it portends politically.

Chile has not participated heavily in the collection and exchange of major crops, as have Mexico, India, and Brazil, but agricultural scientists feel that the IPGRI is now in an excellent position to stimulate and develop programs for minor crops. They also see some progress in the addition of farmers' rights to those of plant breeders in the Undertaking.

Finally, Chile's analysis of the diversity of each species is a task that will involve numerous scientists over several years. If the public sector allocates adequate resources to these conservation efforts, plant breeders as well as genetic engineers will have a wide variety of germplasm to grow and improve upon for the coming years. Chilean crop varieties, whether landraces or breeders' lines, represent an invaluable potential for agricultural research. By improving the regional crop varieties that are already integral parts of the existing agricultural complex, agricultural scientists will contribute to keeping food production stable, thereby diminishing external genetic and technological dependence.

Conclusions

Conservation of the genetic diversity of Chile's crop plants and their wild and weedy relatives involves biological as well as social changes that require studies, training, and explicit action programs. These have much in common with the more visible concerns directed not only at maintaining the diversity of wild species of plants and animals and the natural ecosystems they are part of but also at assuring the future of agricultural productivity and food security.

Budgets to support the necessary research and development of effective germplasm conservation projects are inadequate and must be enlarged. There is growing awareness of the objectives and need for support to accomplish them. The translation of this awareness into needed financial support will have to come primarily from the public sector, a task made more difficult by the free market ideology currently in vogue. The industrial sector will be able to help, as will international agencies. Farmers such as those on the Chiloe Islands also have an important role to play. Once the country has assured the conservation of its genetic material, it will be able to participate in the international race to develop the new biotechnologies, but not before. In any event, the philosophy of germplasm conservation must not be overlooked or forgotten.

PART FOUR

Toward a Global Culture

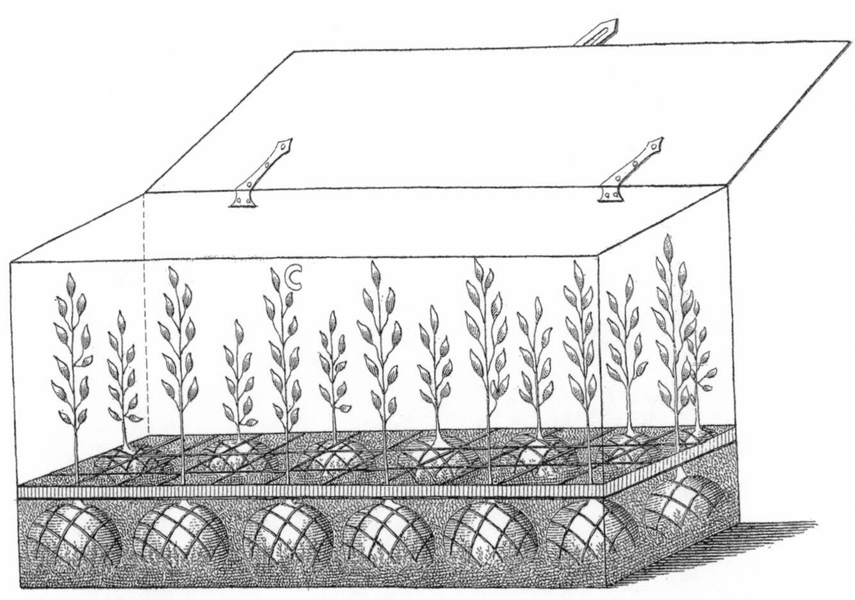

The Inside of the box shewing the manner of securing the roots of W. Florida and W. India plants surrounded with earth & moss tied with packthread and fastend cross & cross with laths or packthread to keep them steady.

10

Cultures of Property

> Imagine the political outcry if a company discovered a useful disease resistance gene in a natural Mexican weed and then sought to patent it in commercial varieties of maize that would be sold back to Mexico.
>
> John Barton

Each of the four nations we have examined has developed somewhat different institutional structures to address the issue of germplasm conservation. At the same time, an international system has been in the making. Furthermore, as in most human affairs, the creation of an international system has involved disputes, conflicts, and heated argument – and the process is far from over. It still continues today and is likely to be shaped further by future events. A central feature of this system for germplasm conservation has been intellectual property rights for plants. It is to these complex legal debates that we now turn.

Intellectual Property Rights and Plants

Until several centuries ago, seeds were not seen as commodities anywhere. A farmer who saw a novel variety or species in a neighbor's field would obtain a few seeds free of charge. No one specialized in the sale of seeds because there was little or no money to be made in it. This situation is still typical in most developing nations. It contrasts markedly with mechanical innovations, which have long been subject to intellectual property protection.

Utility patents, that is, limited grants of rights of property in ideas for certain things and processes, have been available for machines since just after the beginning of the industrial revolution. Utility patents provide monopoly rents to the patent holder for a given time, in principle thereby compensating for the costs of research and development in exchange for public disclosure of the details of the invention. Until recently, utility patents could be obtained only for machinery and certain chemicals; living organisms were excluded. Thus mechanical research became a private sector activity while biological research was taken up by the public sector.

Moreover, in the past seeds have offered a particularly large stumbling block to capital investment. As Buttel and Belsky (1987) have noted, there are three obstacles to profitability in the seed industry: farmers, public sector research, and other seed companies. Moreover, seeds are not only a means of production but a means of *re*production. On the one hand, they can produce a crop; on the other hand, that same crop can be used to provide seeds for the following year. In short, the problem for the entrepreneur with capital to invest was that the seed contains within itself the means for its own reproduction. A machine could always be copied, but the process often was slow and painstaking and required sufficient capital to purchase the equipment necessary to make the machine. In contrast, only a single seed was needed to copy a variety. Hence seed companies began to search for other solutions: changes in the public research agenda, hybrids, plant patents, plant variety protection, and, most recently, utility patents (see table 10.1).

PUBLIC RESEARCH

During the last century most developed nations created public agricultural research systems. Busch and Sachs (1981) estimated that by 1900 nearly 600 agricultural experiment stations existed worldwide and by 1930 more than 1400 could be identified. These research organizations have tended to focus on the development (and often free distribution) of improved varieties of plants, thereby competing directly with private seed companies.

HYBRIDS

Hybrids provided a partial solution to the problem facing the would-be plant breeder. When planted, hybrid seeds will produce an abundant harvest. Harvested seed, however, when planted again, produces a very poor crop. Hence the hybrid serves as a kind of biological patent because it can be copied only by obtaining its parents. Thus farmers who adopt hybrid seeds can no longer compete in seed production. Moreover, the price of varietal seed, where farmers are potential seed producers, can never move very far beyond the cost of on-farm production. With hybrid seed, the price of seeds could be and has been sepa-

Table 10.1 Comparison of Plant Variety Protection, Plant Patents, and Utility Patents for Plants

	Plant Variety Protection	*Plant Patent Act*	*Utility Patent*
Complete written description required	no	no	yes
Sexually reproduced varieties protectable	yes	no	yes
Asexually reproduced varieties protectable	no[a]	yes	yes
Protect hybrids	no	yes	yes
Novelty required	yes	yes	yes
Requires standard of nonobviousness	no	no	yes
Provides generic coverage	no	no	yes
Provides protection for genes and other parts of plants	no	no	yes
Doctrine of equivalents available	unknown	yes	yes
Sexually reproduced varieties infringe	yes	no	yes
Asexually reproduced varieties infringe	yes	yes	yes

[a]Asexual reproduction infringes the certificate except when applicant is in pursuit of a plant patent.

Source: Sidney B. Williams Jr., "Utility Product Patent Protection for Plant Varieties," *Trends in Biotechnology* 4 (February 1986): 34.

the cost of seed production. Because seeds are a relatively small percentage of the overall cost of farm production, farmers willingly pay considerably more for hybrid seed, affording seed companies a substantial profit. In short, protecting the parents from theft affords a virtually foolproof type of "patent" protection.

Nevertheless, hybrids have limitations. Of greatest importance is that hybridization is not an equally simple task for all crops. Maize, the first plant to be commercially hybridized, has the major advantage (from the point of view of the seed producer) that separate male and female flowers are located at different places on the plant. The detasseling of maize is a relatively straightforward process involving simply slicing the top off of each plant. Once this is accomplished, the plant can be easily pollinated by another variety. In contrast, crops such as wheat are usually self-pollinated and are exceedingly difficult to emasculate by hand. In addition, Berlan (1987) has argued that the ratio of seeds planted to seeds harvested is of particular importance. A single maize plant may yield hundreds of seeds; a wheat plant will yield just a few. Obviously, the closer the ratio is to 1.0, the less interest there will be in developing hybrids.

UTILITY PATENTS

Proponents of patent law, as it originally emerged out of the industrial revolution, argued for it on two grounds. First, and most important, it was said that patents created an incentive to invention and to technical progress. Invention was seen as desirable to provide better, more, new, more efficient, and different goods and services to the nation. Second, and largely as a side effect, patents (if sufficiently long term and adequately enforced) permitted inventors to make a considerable profit on their invention. In addition, the act of public disclosure was said to serve several purposes: it would prevent infringement and make the knowledge itself public, thereby (in principle) encouraging further invention (Beier, Crespi, and Straus 1985).

Of course, no one has argued that patents are the only way to stimulate invention. Evenson and Evenson (1983) have suggested several other policy instruments that might be used to accomplish the same goals: monopolistic industrial organizations, prizes for research and development, research cooperatives, and public research institutions.

Nor has there been universal agreement that patents truly serve a social good. To the contrary, they have been the subject of much dispute since their inception. For example, Jefferson could hardly be said to be a supporter of patenting (Bathe and Bathe 1935). Furthermore, one of the few comprehensive studies of a patent system ever conducted (Taylor and Silberston 1973) concluded that only the pharmaceutical industry benefited from additional innovation as a result of patents. Similarly, a survey of research and development executives conducted by Levin et al. (1987) suggested that only pharmaceuticals and perhaps semiconductors relied heavily on patents to promote invention. They also pointed out that a uniform patent law did not necessarily produce uniform results because each industry was affected differently. Several biotechnology patent bills have been defeated in Congress precisely because what is seen as desirable by the chemical industry is seen as equally undesirable by the pharmaceutical industry (Raines 1991–92). Even authors who are strong supporters of expanding the patent system concede the lack of evidence of its social benefits (Beier, Crespi, and Straus 1985).

Before considering the relatively recent changes in patent law with respect to plants, let us consider what, traditionally, as a result of legislative decree or legal precedent, have been the requirements for patents. Though we take the United States as our case in point, the general principles apply to all nations having patent systems, although the details as to what may or may not be patented and the length and scope of the patent vary from nation to nation.

A first criterion for a patent is novelty. For legal purposes, novel means a novel composition of matter, though the things combined need not be novel in and of themselves. Discoveries were normally excluded from this definition on the grounds that they were not created by the discoverer but merely revealed. The requirement of novelty is usually established by demonstrating that the invention in question has not been already patented by someone else and that it is not a common practice of practitioners of a given art or technique.

A second criterion for conventional patents is nonobviousness, that is, the item to be patented could not be easily created by a person with an ordinary skill in the art. For example, if we are producing breakfast cereals, we cannot simply change their shape and obtain a patent. The novelty we create must involve something more fundamental. Unlike the criterion of novelty, however, nonobviousness is more difficult to prove. What appears obvious after the fact may have appeared impossible beforehand. Moreover, no simple test for nonobviousness can be applied, as in the case of novelty, because there is no record of what an ordinary practitioner in a given field might find obvious (Emery 1987). In addition, the effort involved is *not* a factor in determining novelty or nonobviousness. Something that requires great effort to create but is produced using conventional techniques known to the ordinary skilled practitioner has not traditionally been considered patentable.

The third criterion for utility patents is utility. The patent applicant must demonstrate that some public benefit will derive from the invention. This criterion has been used to exclude certain inventions from patenting on the grounds that they are harmful to the public good or useless.

The final criterion for utility patents is specification. A written description of the invention that may be used by a person with ordinary skills in the field must be filed. Such specifications would ordinarily permit others to use the invented product or process, justifying the patent system by ensuring that knowledge of the novelty would be in the public domain. The specification is also used by patent examiners to determine if a patent should be granted.

Traditionally, certain items have been excluded from utility patent systems. For example, basic research findings have been excluded from the patent system (though the new biotechnologies blur that distinction). All life forms, including plants, animals, and microorganisms, have also been excluded from patent protection. The grounds for exclusion are lack of invention, lack of novelty, the sacredness of life, and the impossibility of providing an adequate specification of the life form. Let us briefly consider each of them.

The first is that the creation of "new" living things by people involves not an invention but a discovery. That is, what is being done in, for example, plant breeding, is not the invention of a new organism but the discovery of an organism or variety that has certain particularly desirable properties. In conventional breeding, the methods are well established and the material is well known; what is novel is the particular combination of genes that the breeder has managed to put together, through a good deal of hard work, in a given variety. Moreover, the breeder, more likely than not, is unable to explain how the variety was created. Hence there is no clear *idea* to patent; instead, there is a concrete object called a variety. Thus, using this argument, plants were excluded because they were not considered to be cases of invention.

A second argument, not unrelated to the first, is that living things come in an infinite variety and that there is no way of knowing if nature has not somewhere duplicated exactly the work of the plant breeder. This is quite different from machines, which are not found in nature. Therefore, it has been argued that there is no way to know if a variety is truly novel.

A third argument hinges on the ethics of patenting life forms. The argument is that life is sacrosanct and should never be considered the property of anyone. Although this argument has been used by opponents of plant patenting, it has (arguably) been of less importance in excluding life forms from the patent system than the two other reasons listed above.

Finally, it is argued that no adequate specification of life forms could be provided. Although the item under consideration might be described, merely describing it would be inadequate for the purpose of reproducing it. Put differently, even the most detailed botanical or genetic description of a plant will not permit its creation. This objection has been sidestepped for both microorganisms and plants by deposit of samples of the material in special collections held for that purpose in addition to the written description (Jeffrey and Schlosser 1990).

Scientific data have traditionally been excluded from patenting on the grounds that they are discoveries rather than inventions, there is no way of establishing that the same process has not occurred somewhere in nature, and knowledge should circulate freely so that science can progress.

PLANT PATENTS

In the United States, the first patentlike protection offered for plants was embodied in the Plant Patent Act (PPA) of 1930, although the act was limited in its coverage to asexually propagated plants (e.g., many flowers and tree crops that can be grafted rather than grown from seed) because it was felt that the genetic variation in sexually propagated plants would make it impossible to enforce patent claims. Moreover, unlike utility patents, holders of plant patents are not required to explain how to produce the desired product. Descriptions needed only to be "as complete as possible." Obviously, for many products of asexual reproduction, it is virtually impossible to reproduce the process whereby they are created. Thus patent holders are required instead to place a sample on deposit in a public repository.

In addition, plant patents are narrower in scope than utility patents in several ways. First, the patent can include only one claim directed to the variety. Second, a given patented plant may be grown outside the United States and its fruits imported and sold here. In contrast, a patented machine produced outside the United States could not be imported without infringing the rights of the patent holder. Thus the Plant Patent Act gives considerably less protection to invention than do utility patents.

PLANT VARIETY PROTECTION (PVP)

It was not until some 30 years later that the idea of patentlike protection for other plants became an issue. It came to be known by the somewhat specious title of plant breeders' rights (PBR). The argument advanced by proponents of PBR was that compensation was needed for the considerable investment made to develop new plant varieties and that plant breeders should have the right to that compensation. The argument was partially specious in several ways. First, it implied that the typical plant breeder was self-employed or at least in some way at risk if some compensation was not provided. Second, it assumed that the typical breeder worked for the private sector instead of for some public agency. The first implication of the term was and remains inaccurate; most breeders today work for large-scale organizations and are themselves simply salaried employees. The second has become accurate in large part as a result of the passage of PBR legislation; it was certainly not the case in the 1950s when the idea was initially proposed.

The Paris Conventions of 1883 and 1925 as well as the Act of London of 1934 of the Paris Convention all include agricultural products within their definition of industrial property. Nevertheless, various practical and legal considerations barred the granting of utility patents for plants. The idea of plant breeders' rights was first proposed in 1956 at a seed industry conference. In 1957 the French government convened a conference of agricultural officials to discuss the issue. As Bent et al. (1987:52) note, "It had been the agriculture ministries that had generally opposed the patenting of plant-relating inventions in the decades preceding the 1950s." In short, developing separate PBR legislation had the effect, desirable from the point of view of the agricultural interests involved, of ensuring that industrial and agricultural property protection would take separate paths.[1] The convention was signed in Paris on 2 December 1961 by Belgium, Denmark, France, the Federal Republic of Germany, Netherlands, Italy, and the United Kingdom. It was ratified by 1968. It established the Union pour la Protection des Obtentions Végétales (UPOV). The signatories could not agree on which crops would be covered by the convention so it was decided to let each nation make that decision on its own, though it was agreed that the convention should work toward inclusive coverage of all crop plants. As a result, lists of protected species vary from nation to nation, but the details of the law are essentially the same in each nation.

Fejer (1966:4), in an article in a semipopular journal published by the USDA, argued that most public sector breeders opposed plant variety protection. He also predicted that "if international considerations force the introduction of breeders' rights on the United States, it is predictable that government institutes will gradually withdraw from breeding such crops, and confine their activity to basic research." As we shall see, this is very much what has been happening, though perhaps for some reasons that Fejer could not have foreseen in 1966.

At the same time, the UPOV convention and the individual national acts that followed it served to distinguish the seed companies and their allies in the ministries of agriculture of the world from the industrial firms and their allies in patent offices. Because varieties registered under PBR legislation were to be deposited and tested by government agencies, and only ministries of agriculture had such facilities at their disposal, the PBR legislation also reflected a victory of agricultural over industrial interests.

The U.S. Plant Variety Protection Act (PVPA) of 1970 and similar acts in Europe permitted for the first time a patentlike protection for sexually propagated plants. (It simultaneously prohibited registration of asexually propagated materials on the grounds that the PPA already covered them.) The criteria for variety protection certificates are considerably less stringent and the protection afforded certificate holders is somewhat less than those for utility patents. Under the terms of PVPA a certificate of protection can be granted if a plant satisfies three criteria: novelty, uniformity, and stability. The first of these criteria required that all registered plants be distinguishable from existing varieties. This interpretation of novelty has four peculiar characteristics (as compared to utility patents): It does not limit novelty to useful differences, that is, those of interest to farmers and processors. It glosses over the essentially impossible task of discerning whether a new variety is in fact new in the sense that it does not exist somewhere in nature or even in some farmer's field. It does not require that someone with ordinary skill in the art of breeding not be able to produce the variety. Finally, it refers to the concrete object – the variety – and not to an idea.

The second criterion for registration under PVPA is uniformity. Only a variety that uniformly expresses a novel characteristic is eligible for a certificate. The uniformity desired consists in the identity of all plants of the certified variety. Nothing about the uniformity is necessarily of interest to farmers. Indeed, a variety that is agronomically desirable but not uniform is not eligible for a PVP certificate. Finally, the novelty must be stable; it must be a permanent feature of the variety over several generations.

In return for meeting these requirements, the certificate holder can bar unlicensed parties from marketing the seed for the protected plant for a period of 20 years. Unlike utility patents, however, PVP certificates provide no control over either the use of the protected variety for future research or the multiplication of the seed by farmers for their own use and even for sale as long as they use no promotion or advertising. Moreover, it is only the variety that is protected under PVPA; unlike utility patents, no discretion with respect to scope is granted to the applicants or examiners.

One of the key provisions in the U.S. law restricts plant variety owners to selling their protected varieties as a class of certified seed under Title V of the Federal Seed Act of 1938. This provision has led the Crop Improvement Association leaders to endorse PVP as way of protecting and encouraging investment by private firms in research that would otherwise not be conducted (Michigan Crop Improvement Association 1990). This view is held by seed breeders in private corporations as well (Schillinger, personal communication). The growth of private research in soybeans over the past decade has been used to build their case for plant variety protection, although some plant breeders view the introduction of PVP as an attack on the role of the government in research and on the "public spirit" of plant research in general (Freed, personal communication).

The granting of PVPA certificates did have the expected effect of encouraging greater private sector investment in plant breeding. The bulk of the interest, however, has focused on several crops. Soybeans and wheat account for about one-third of the U.S. certificates issued. Nor is there any clear evidence that the overall private investment in plant breeding has increased much (CAST 1985). At the same time, public investment has declined. Moreover, the act had the effect of increasing the value of existing seed companies, many of which have now been bought by the large chemical and pharmaceutical companies. Plant variety protection was designed with traditional breeding in mind, however, not with the new techniques introduced later. In 1991 the UPOV Convention was revised (UPOV 1991). Several major changes were made including extension of varietal protection to those "varieties essentially derived from" the protected variety, rewording the farmers' exemption as a voluntary rather than mandatory procedure, extension of the protection to 20 years for plants and 25 years for trees and vines, and eliminating the clause prohibiting double protection. Perhaps the most important of the changes is the first because it goes to the central reason behind PVP. Unlike mechanical invention, which had been characterized (perhaps wrongly) as the individual pursuit of a novel idea that is materialized in a particular invention, plant breeding involves a novel combination of known characters using well-established techniques. Thus plant breeding is much more a collective process in which the work of each breeder directly builds on that of others and the result is always – in principle – obtainable by others equally skilled in the art. This is apparent in the results obtained. For example, Hermitte and Joly (1991) note that of 136 cultivars of winter wheat in France listed in the catalog, only 12 lacked a common parent with the others. Moreover, the five most important parents were found in 75 varieties. Clearly, the way in which "varieties essentially derived from" is interpreted will have a significant effect on the practical results obtained.

These changes have the collective effect of making PVP look much more like utility patents. Furthermore, some have argued for the elimination of the research exemption in the UPOV Convention. Yet, ironically, in so doing they undermine the very logic used in developing a distinct PVP system initially. Finally, although many nations subscribe to the Paris Convention, far fewer are signatories to the UPOV Convention.

TRADE SECRETS

Yet another way of protecting intellectual property is through trade secrets. A trade secret is information necessary to the commercial viability of a particular product or process that is not subject to patenting but is held privately. The Uniform Trade Secrets Act (14 U.L.A. 537–51 [1980 and Supp. 1986]) defines the theft of trade secrets as a tort. Other nations have similar laws and civil and criminal penalties. Trade secrets are not the same as ordinary secrets. Thus one might keep some piece of basic research secret, but the penalties of trade secret laws could not be invoked if someone were to copy the information and make it public. Of key import in trade secrets is the commercial advantage they confer. Basic research results are usually not held to confer such an advantage (Eisenberg 1987).

Moreover, unlike patents, trade secrets may be maintained indefinitely. Trade secrets, however, are not violated by independent invention. In other words, if a seed company keeps its novel breeding process a secret, but another company independently invents the same or similar process, the trade secret statutes are of no value. This is in marked contrast to utility patents, which are infringed upon even by independent invention. To date, relatively few trade secrets exist related to plant improvement, but the new biotechnologies certainly offer that possibility. Furthermore, unlike approaches to breeding, many biotechnological secrets would be relatively difficult to invent independently (e.g., gene transfer strategies). Thus trade secrets might be particularly desirable in cases where patent protection is narrowly defined, making it easy for competitors to design around the patented process or product. In such cases, patenting would be tantamount to making the invention available for free public use.

French intellectual property rights are essentially the same as those that apply in the United States. One notable exception, however, is the peculiarly French concept of dependent licensing. This differs from the American notion of compulsory licensing in that it recognizes explicitly those cases in which someone may make an improvement on an existing invention. In those cases, cross-licenses may be required by the court. In other words, both the holders of the original patent and the holders of the patent on the improvement would be licensed to manufacture and sell each others' product. Such court decisions are relatively rare because they come into play only when the two parties are unable to reach an agreement without court intervention. As we shall see, the dependency concept has potential application in the area of plant improvement (Hermitte and Joly 1991).

Brazil was one of the original signers of the Paris Convention and has had a utility patent system since that time. Brazil, however, did not sign the more recent Stockholm text. The Industrial Property Code is administered by the National Industrial Property Institute (INPI), which has a staff of about 100. The system is somewhat uneven in its coverage, and pharmaceuticals and food are excluded. The legal process is time-consuming and the penalties are weak. Trade secret legislation exists but is weak (Frischtak 1990). There are no plant variety protection laws. For these reasons Brazil is one of only four nations found on all five lists of "problem" countries identified by U.S. trade associations (Rozek 1990).

The advent of the new biotechnologies has heightened interest in intellectual property in Brazil. For example, Biomatrix, a small high-technology subsidiary of Brazil's largest seed company, Agroceres, has joined 33 other biotechnology companies in ABRABI, the Associação Brasiliera de Empresas da Biotecnologia, to lobby for patent protection for biotechnological inventions (Grynszpan 1990).

In short, the competition-cooperation paradox is visible here as well: "The Brazilian government, therefore, perceives itself as being faced with the choice of either altering its long-standing position on intellectual property protection and risking its credibility as a leader of developing nations on the one hand, or facing the uncertainty posed by potentially escalating U.S. retaliation on the other" (Richards 1988:165). Recently, Brazil caved in to the pressure applied by the West and began to draft a new intellectual property law. The final draft of the new National Code on Industrial Property, which includes biotechnology and pharmaceuticals, was presented in April 1991. It provides a 20-year term with five extra years for local manufacture. Legislation providing for PVP is also being drafted (Biotechnology and Development Monitor 1991). Most academics oppose the law, but by June 1993 it had been approved by the Chamber of Deputies, the lower house in Brazil's parliament (Hart 1993).

In Chile biological methods may be patented. In addition, seeds that are new, stable, and homogeneous may be registered (Bent et al. 1987). Like Brazil, however, Chile does not subscribe to the UPOV convention, although UPOV has encouraged Chile to join. While the Chilean government is as yet uninterested, some politicians have urged that plant variety protection be extended to grapes, berries, and other fruits that are sold in the profitable counterseasonal export market.

RECENT DEVELOPMENTS

A key case in the application of utility patent laws to life forms was the now famous 1980 Supreme Court decision in *Diamond* v. *Chakrabarty* (447 U.S. 303). Though the specifics of the new case involved a microorganism, the decision opened the door to patents for all life forms. Moreover, it eliminated the argument about novelty that always excluded such patents because it involved the actual reconstruction of a microorganism using recombinant techniques. In that case the government argued that living things were treated specially by Congress in both PPA and PVPA and that genetic technology was new and could not have been foreseen by Congress. The court argued in turn that the issue was whether the product was human-made and that not allowing the unanticipated invention conflicted with the entire premise of patent law (Linck 1985).

Following the decision, the Patent Office began to issue patents for plants. At the same time, the U.S. government reversed a long-standing position and in 1980 permitted universities to patent inventions created as a result of government grants. This had the effect of creating a new constituency in the financial offices of research universities. Because the majority of universities are under severe fiscal constraints, patents for life forms were gleefully received by university officials. Then, unexpectedly, in 1984, the Patent and Trademark Office reversed itself and argued that because plants were specifically covered by PPA and PVPA, coverage by utility patents was inappropriate. Then, in 1985, as a result of *Ex Parte Hibberd* (227 USPQ (BNA) 443) the U.S. Patent Office finally reversed over 100 years of precedent and began to permit utility patents for plants.

Although the European situation has been less dramatic than that of the United States, there have been several key cases in Europe as well. First, in a 1983 decision of the Technical Board of Appeals of the European Patent Office (decision T59/83, Propagating material/Ciba-Geigy, 26 July 1983), it was determined that anything that was not a plant variety in the sense of UPOV was patentable. In a later decision (T320/87, 10 November 1988), the board argued that a process for breeding hybrid plants was essentially nonbiological and therefore patentable. Moreover, the latter case shifted the court practice from determining what was microbiological to determining what was essentially nonbiological. Thus, anything essentially nonbiological could be patented (Hermitte and Joly 1991). Indeed, in March 1994 the European Patent Office issued a patent on all transgenic varieties of soybeans to Agracetus, a subsidiary of W. R. Grace – a patent that is sure to be hotly contested. These changes in the European Community have been matched by a tendency on the part of the French courts to consider effort as an acceptable substitute for inventive activity (Hermitte and Joly 1991).

To understand why the U.S. Patent Office and the European Patent Office should make such changes in policy, it is necessary to consider just what the new biotechnologies can do that conventional plant breeding cannot. These new technologies, consisting of techniques ranging from tissue culture to recombinant DNA, have considerably expanded the options open to plant scientists. They have introduced new precision and dramatically accelerated the processes of plant improvement. In so doing, they have shifted attention from a macroscopic stochastic understanding of a plant to a microscopic deterministic approach (Hermitte and Joly 1991). Hence these new cellular and molecular techniques have served to transform the horizon of possibilities within which both science and business operate. Biotechnology is responsible for three interrelated changes. First, it has challenged the time-honored belief that plants are products of nature, somehow different from human-created products. Second, it has served to increase interest in the conventional area of plant research. Finally, it consists of tools that on the one hand serve to advance science and on the other hand make patent enforcement possible in ways never before possible. Let us consider each of these in turn.

First, common sense and patent law have respected a distinction between inventions – products of human ingenuity – and discoveries – the revealing of aspects of nature. Inanimate things could be inventions (e.g., pesticides, tractors, computers), while animate things could only be discovered. The new biotechnologies make new distinctions by making possible the transfer of genetic material outside of the normal processes of sexual reproduction. Many legal scholars have argued that utility patents should be extended to plants because the exclusion of plants from patenting was what philosophers call a category mistake on the part of legislators and the courts (e.g., Krosin 1985). They argue that the legislatures and courts erred in assuming that all things animate were also by definition "products of nature." Yet this position does little to explain why the patent offices of dozens of nations should have excluded plants (and animals) from patent protection for more than a century. Surely the patent officials and legislatures were not *all* blind.

Another hypothesis appears far more plausible. Before the advent of the new biotechnologies, the legislatures and courts did not and need not have worried about the implications of treating all animate things as products of nature. Only after the creation of the new biotechnologies did the categories "animate" and "product of nature" emerge as distinct. For example, scientists have performed many technical feats recently. Using the new technologies, genes from any source may – at least in principle – be inserted into any host. It appears that plant scientists may now create wholly new life forms, rather than being limited to the manipulation of old ones. For example, *Science* reports tobacco with human antibodies (for testing for human diseases), introduction of an enzyme from carrots into soybeans to increase oil production, and insertion of an antifreeze gene from fish into tomatoes and tobacco (Moffat 1991). Clearly, none of these new genetic combinations could have occurred in natural settings. Nor could they have occurred in scientific laboratories as little as 20 years ago.

At the same time, the new technologies blur old distinctions such as that between microbiological and "essentially biological" processes. In times past, microbiologists studied and manipulated microorganisms while macrobiologists worked with higher organisms and parts thereof. When molecular biological techniques were applied to plants and animals, it became impossible to distinguish between micro and macro processes. Again, this development could not possibly have been foreseen by either legislators or the courts as little as 25 years ago. For example, recently, the European Patent Office decided to grant a patent for modified alfalfa cells. The Swedish Centre Party challenged the patent. As its leader, Ola Jonssen, argues, "We feel these things should be discussed openly and decided democratically, not left to technical processes inside the EPO" (Kingman 1989:10).

In the *Chakrabarty* case, the Supreme Court went even further in arguing that not allowing unanticipated invention contradicted the position of Congress in developing a patent system. Yet the court seems confused here as well, conflating the new biotechnologies – a class of techniques – with a specific process. The issue is that Congress treated life forms differently than inanimate objects for all inventions in those classes.

In short, many legal scholars – including the members of the U.S. Supreme Court – in their attempts to reinterpret patent law in the light of new technologies read the present into the past. The problems that confront us today as a result of technical changes (themselves the result of needs and desires of various groups in society) have been read into the record of past decisions. Not surprisingly, when this has been done, the record of excluding plants from patenting has been found wanting. Yet we must recognize that the source of the difficulties in patent law lie not within the law itself but within the changing technical horizon of biology.

Second, the very power of the new biotechnologies has had the effect of increasing corporate interest in patenting plants and plant materials. It appears clear that a wide range of new products may be derived from these new techniques. These include

1. secondary plant metabolites produced by putting plant cells in bioreactors and treating them as if they were microorganisms. Such biochemicals are difficult to produce and often expensive.
2. insertion of (parts of) microorganisms in plants. For example, the toxin from *Bacillus thurengiensis* has been used to create insect resistance in numerous plants.
3. selection of plants for chemical tolerance. This would permit the use of herbicides after emergence of the crop plant in the field.
4. more rapid testing of exotic materials for resistances. For example, thousands of maize varieties might be tested for salt tolerance by growing tissue samples in petri dishes.

Together these new opportunities represent a considerable profit potential for the chemical, pharmaceutical, seed, and food companies. Not surprisingly, these companies desire to expand the scope and geographical coverage of patent protection so as to recover their considerable investment costs, increase their profit margins, and increase their market shares. Moreover, unlike the situation as little as 25 years ago, the seed industry is now dominated by large agribusiness companies that operate worldwide rather than locally or regionally. Not surprisingly, representatives of these companies argue that patent laws should abandon the distinction between animate and inanimate matter (Beier, Crespi, and Straus 1985). This is quite distinct from general popular opinion that sees patenting of life forms as immoral and unethical (Lacy, Busch, and Lacy 1991).

Moreover, many public breeders in each of the countries we examined are opposed to both PVP and patents. As one Brazilian breeder at EMBRAPA said, "It is not just the plant breeder who produces new cultivars; behind him/her there are an uncountable number of other professionals involved in the process without whom work would be practically impossible. Why should we reward only the plant breeder?" Similarly, owners of small independent seed companies worry about patents. The owner of a French seed company put it this way: "Big multinationals are pushing for patenting because they hope to control the market through it. If a company finds a resistance gene and incorporates it into an agronomically useful variety, then it will demand a royalty. We believe that what is living is not patentable." Yet in the current political milieu, it is unlikely that these voices will be heard.

Third, the new technologies are changing the definition of variety. In the past, it was possible to determine that two plants were of the same variety only by phenotypical comparisons. Two plants with different genotypes might appear phenotypically the same so the uniqueness criterion could be enforced only at the level of the whole plant. Current techniques, including the use of restriction fragment length polymorphisms (RFLPs) and polymerase chain reactions (PCRs) permit the documentation of a "molecular signature" that is unique to each variety. Thus in applying for a utility patent, the genetic map of the new cultivar can be offered as evidence of its uniqueness.

The new sword cuts both ways, however. By redefining "variety" to include a uniform genetic code only, many breeders fear that we will open a Pandora's Box. The variety pirates will merely inject a small (and purposefully innocuous) string of genetic material and claim that they have created a new variety (Schillinger, personal communication). And who is to say how much difference at the genetic level is sufficient to indicate novelty?

The new biotechnologies have also contributed to the restructuring of public sector research that started with the advent of plant patents in 1930. Specifically, at each stage in the development of greater proprietary protection, the division of labor between the public and private sectors has changed. Public research has retreated toward the more basic – read, less profitable – end of the research spectrum. In the United States and France public sector research organizations have eliminated much of the breeding of finished plants. In Britain the once world-famous Plant Breeding Institute at Cambridge has been sold to the private sector. At the same time, the seed industry has become more concentrated, more competitive, and more international in its scope (Buttel and Belsky 1987; Kloppenburg 1988).

In 1989 USDA sponsored an invitation-only workshop on the problems of plant patenting attended by representatives of the seed industry and the scientific community. Although no overall consensus was reached at the meeting, there was agreement on the desire to eliminate the farmers' exemption in PVP and to expand the research exemption on utility patents of plants. Participants were also concerned about the patenting of traits as opposed to genes. Moreover, "Germplasm exchange within the United States has decreased since the PVPA was passed in 1970, as well as since 1985 when utility patents [f]or plant materials were first allowed, workshop participants acknowledged" (*Diversity* 1989:37).

Similarly, the American Seed Trade Association (ASTA), after surveying 125 of its members, supported patenting of genetic components of plants but PVP for varieties. It supported eliminating "near duplicate" varieties. It also announced its support for compulsory licensing and, surprisingly, farmers' rights to save seed (*Diversity* 1988a).

Despite the advent of PVP and the *Hibberd* decision, USDA policy is to continue to release varieties on a royalty-free and nonexclusive basis unless it appears more desirable for U.S. agriculture to obtain a PVP certificate (*Diversity* 1988b). The various state experiment stations have adopted a wide range of positions on the issue.

In Europe the International Chamber of Commerce and the International Association for Industrial Patenting have lobbied for the extension of patent rights to plants. The industrial view is perhaps best expressed by John Duesing (1989:23): "It is Ciba-Geigy's position that legal protection of intellectual property serves the public interest by stimulating continuing investment in technological innovation. The protection must apply to all scientific advances, including those from the areas of biology and plant breeding." Duesing goes on to note that the company also opposes compulsory licensing, which it sees as leading to genetic uniformity.

In contrast, the Committee of Professional Agricultural Organizations (COPA) and General Committee of Agricultural Cooperatives (COGECA) opposed patents (Comte 1989). The International Coalition for Development Action, a European nonprofit organization, has also opposed patenting. Similarly, at least some plant breeders see the move to permit patenting as wholly undesirable and encouraging monopoly in the seed industry (Hardon 1989). The two sides in the dispute are reflected within the Directorates of the European Community: DG III (industry) has supported plant patents, while DG IV (agriculture) has opposed it (Hermitte and Joly 1991).

Of greater import for us here, however, are the complex ethical problems associated with the patenting of life forms and particularly those of relevance to the conservation of crop plant germplasm. There are three of specific concern: the new definition of novelty, the problem of disclosure, and the issue of genetic erosion.

Novelty in patent law has always meant that the item for which a patent was claimed was significantly different from any other item ever patented or in general use. Extension of the concept of novelty to the living world raises a series of new and complex questions. Auerbach (1983) suggests that the criterion of novelty can be met by establishing that others had neither known of nor used the material in question or patented it elsewhere. In contrast, Crespi (1988) has argued that novelty in patent law relates only to what was known before, not what existed before. Yet the notion of a variety is ambiguous (Berlan and Lewontin 1986b). The highly trained eye of a plant breeder is needed to discern the difference between two varieties – or at least it did until recently, when biotechnology began to narrow the definition of a plant variety to a unique genetic code. Yet such a narrow definition, though clearly more precise than those used in the past, confers no necessary or desirable agronomic, processor, or consumer differences on the final product.

The problem of novelty is a two-edged sword. If something is to be considered novel by virtue of the specific gene sequences incorporated in it, then novelty should be easy to prove. At the same time, however, other varieties with similar but not quite identical sequences should also be patentable in their own right. These differences may have no agronomic or other relevant value but still permit the applicant to claim novelty (as well as nonobviousness). If interpreted in this way, the patent might prove virtually useless. Yet if novelty is interpreted more broadly so as to include only useful properties, it would be more difficult to prove, but potential competitors would be shut out of the market more easily.

This same problem exists within the framework of Plant Variety Protection. Many complaints have been received about "new" varieties that differ from existing varieties merely in some trivial (but distinct and stable) character. This has become known in the seed trade as cosmetic breeding. The revised UPOV Convention excludes "varieties which are essentially derived from the protected variety" (UPOV 1991:21). It is likely that RFLPs will be used to determine the similarity among varieties, but no definition of "essentially derived" has yet been promulgated. Clearly, if it is defined broadly, it would extend the protection considerably and block competitors in much the same way that utility patents do. If defined narrowly, its effects will be minimal (Joly and Trommetter 1991).

The second problem is that of disclosure. Biotechnology threatens to make disclosure impossible by virtue of the sheer complexity of what is being described. For example, Beier, Crespi, and Straus (1985) note that some patents of this type now require over 1000 pages of text to describe. Moreover, the inadequacy of this description is implied by the fact that most nations require (for microbial patents) that a sample of the material be placed in a public collection for some time so that it is available for others to examine. Yet, without an adequate description or a viable sample on desposit, one of the major benefits claimed for the patent system is lost.

Another problem concerns the desirability of double protection. Under the current legal regime in the United States (and possibly but not likely in Europe), it is possible to receive both a utility patent and a plant patent or PVP certificate for the same plant. Moreover, the *Hibberd* decision suggests that the U.S. courts see nothing in the way of double patenting. The recent changes in UPOV open the door to double protection. Straus (1987), in a paper that focuses more on the European situation, argues that PVP certificates might be obtained for whole plants and European Patent Convention patents for plant parts and cells. Clearly, the final word on double patenting is not yet available. Still, one must ask whether the public good is served by such double protection.

Moreover, if plant parts and cells are protected by utility patents and plant varieties by PVP certificates, it is not at all clear which would have precedence. Perhaps a patented gene inserted in a protected variety would be freely available for all to use in developing future varieties. Alternatively, perhaps the variety could not be used for further development unless the gene were removed. Barring a new statute for intellectual property in plants, these issues will be decided in the United States by prolonged and complex judicial proceedings.

Yet another problem is that "patenting will thwart the Congressional intent of making germplasm freely available even though it may have the effect of producing more technological advances in germplasm science" (Adler 1986:207–8). In the PVPA Congress expressly maintained exemptions for farmers and researchers; utility patents grant no such exemptions and thus pose special problems for conservation. Moreover, at least some legal scholars believe that a utility patent would apply not only to the seed but to the harvest (Crespi 1988).

Patents may also encourage monopolization of the marketplace. A handful of firms could wrest control over the world market for plant varieties, especially if the patents granted are broad in scope. That, in turn, could actually slow the pace of technical change because the firms would have little incentive to license their technologies or to engage in further technical change (Mody 1990). A historical analysis of the role of patenting at American Telephone and Telegraph and General Electric (Reich 1985) shows how these companies effectively used patents to block competitors from market entry as well as to steer technical change in directions that suited corporate management. If utility patents for plants are equally broadly conceived, they could similarly restructure the seed industry.

Conversely, even though patents may be used by multinational corporations to block local competition, the absence of such corporations does not necessarily resolve the problem. Goldstein (1988) notes that the Argentine public sector creates parental lines for hybrids and gives them to seed companies at no charge. Moreover, in most developing countries, either there is no private seed industry or it has a local monopoly on seed sales.

Counterposed to these arguments are those of Duvick and Brown (1989). They assert that the situation with patents should not differ greatly from that already present with hybrid maize. For that crop, each company has been forced to develop its own proprietary – and therefore distinct – germplasm pool. Furthermore, they argue that the Third World would benefit from intellectual property laws through rising levels of investment. Yet, though their argument is worth further study, it ignores the very important dependence on the public sector for hybrid maize development. Indeed, the 1970 corn blight was caused by widespread incorporation of cytoplasmic male sterility into extant varieties – a trait developed largely by the public sector. Moreover, the Third World has been virtually ignored by the large hybrid seed companies, despite the patentlike protection that hybrids afford.[2]

The final problem stems in part from the first two and is central to our concerns here. It is, as Fejer noted in 1966, that proof of novelty requires uniformity. This may well have the effect of increasing genetic erosion and certainly will increase the vulnerability of crops in the field. The demonstration of novelty by criteria that are unrelated to agronomic performance, storage, transport, processing, or consumer preference adds a new strain on an already fragile system. And biotechnology is implicated in several ways. It provides the impetus to increase the scope of patentlike protections for plants. It is also through biotechnology that the new procedures for guaranteeing the authenticity of the patent were produced.

In short, despite a large literature that is inconclusive about the claimed advantages for patents in general and serious problems raised specifically by the patenting of life forms, the thrust of law in the United States and other nations is toward an expansion of the patent system to include life forms. In all of the discussions, perhaps the most peculiar document to emerge is the Organisation for Economic Cooperation and Development (OECD) study of biotechnology and patent law by Beier, Crespi, and Straus (1985). The study is symptomatic of the headlong rush to make biotechnology profitable without knowing whether it serves the public good. It is a report of a questionnaire sent to 24 nations as well as interested parties (i.e., the industries affected) requesting information on their patent system for biotechnology innovations as it currently exists and as they would like it to be. The study was accomplished without asking the obvious questions first: Does the extension of patent protection to life forms serve the public good? If not, why not? If so, how would it best do so? Indeed, the authors, either by their own initiative or directive from OECD, avoided the key questions. Instead, they merely state that "there is a need for reliable legal protection on an international scale, if not for additional or stronger measures of international harmonization of patent protection" (1985:11). The authors may be correct. Yet Berlan and Lewontin (1986b:788) have argued that "limiting the use of a good available in limitless quantities at no cost will not be socially useful, will limit the full use of biological potential only to what is patentable, will erect barriers to entry in branches of production where competition is necessary and will limit the free exchange of information between scientists so crucial to science." It is tragic that an organization whose members include most of the world's wealthy democracies has steadfastly avoided these issues and in so doing conceded the issue to those whose short-term profits are increased by such laws.

At least in the United States, there has been a gradual but marked reinterpretation of law over two centuries so as to encourage development interests (Horwitz 1977). Moreover, "In recent years courts have taken a more liberal view of rights of intellectual property holders. . . . The CAFC [Court of Appeals for the Federal Circuit] has also been basically pro-patent . . . ; the CAFC has lowered the standards required to establish novelty. The CAFC also has awarded high damages to patent holders. It has also shown a willingness to permit criminal rather than civil cases against intellectual property offenders" (Mody 1990:215).

Furthermore, because the United States is the world's largest producer of intellectual property, the implications of this position transcend its national borders. For example, the U.S. International Trade Commission has argued that American manufacturers lose $40 billion per year because of pirated ideas.[3] In an attempt to require everyone to adhere to the U.S. system, the country has construed very broadly Section 337 of the Tariff Act of 1930, which attempts to protect U.S. industry from unfair competition. As such, the United States has entered into bilateral negotiations with some 60, mostly developing, nations, each of which it has accused of piracy. When the United States has been unable to get its version of intellectual property rules enforced, it has imposed substantial penalties. For example, it has imposed 100 percent tariffs on $39 million of Brazilian exports (Mody 1990).

The issue of intellectual property rights has risen again in the most unlikely of places: the General Agreement on Tariffs and Trade. In response to private sector lobbying and growing concerns about the size of the trade deficit (Gadbaw and Gwynn 1988), the United States has argued that the lack of adequate – read American-style – intellectual property rights is a nontariff trade barrier. Moreover, although the average patent application takes 18.4 months to process, the average biotechnology patent takes at least 26 months, in part because the turnover rate among biotechnology patent examiners is 20 to 30 percent per year. And small companies cannot afford the legal fees to challenge patents in U.S. courts. Finally, the Bush administration endeavored to make the Patent and Trademark Office self-sufficient by raising user fees considerably (Marshall 1991).

In addition, the United States wants all nations to adhere to an essentially identical patent system that presumably follows U.S. norms. Normally, such a debate would occur within the World Intellectual Property Organization (WIPO), where each member nation has a nearly equal voice. By shifting the scene of the debate to GATT, the United States gains a considerable advantage. The largest trading nations have considerably more weight than others. Finally, the cost of fighting patent infringement suits in the United States is high (Hermitte and Joly 1991). For those few inventors in developing nations, the costs of filing and protecting against infringement are essentially prohibitive.

The irony of all this is that the United States itself depended greatly on pirated technology in its first half century. The story of the theft of plans for spinning mills from Britain is well documented (e.g., Jeremy 1977; Rivard 1974). This piracy permitted the rapid growth of the American textile industry and laid the foundation for U.S. industrialization (Tucker 1984).

Shifting the debate to trade negotiations may also open other issues that the United States would prefer to leave alone. For example, a group of 14 nations (including Brazil and Chile) proposed excluding plant varieties, essentially biological processes, and substances that already exist in nature from patent protection. In addition, some farmers' rights advocates want to include indigenous knowledge in future trade negotiations. Furthermore, many less developed countries have begun to complain of U.S. textile import policies in hopes of striking a deal on textiles for intellectual property rights. Finally, UNESCO has proposed turning its Model Law on Protection of Expressions of Folklore into an international convention. It would include folk inventions such as landraces and medicinal plants (*Diversity* 1990). This is not a trivial issue. One representative of a nongovernmental organization recounted the following story to us: "One family was growing teosinte [a wild relative of maize] in Mexico. Plant explorers brought back a handful. The value was 2–5 billion dollars, because of the disease resistance it contained. The family who was growing it didn't get a penny. If that's not unfair, what is? How to deal with this I don't know." Such stories, though not well documented, are not uncommon. Nor is it clear just who, if anyone, should share the benefits of such landraces and wild relatives maintained for thousands of years by human effort.

Conclusions

Let us summarize the situation.

All nations are dependent on all others for germplasm because the crops they grow include many that originated outside the country. These dependencies date from hundreds or thousands of years ago. Moreover, the developed nations, North America and Australia in particular, have benefited far more from the exchange than have other regions of the world.

Exotic germplasm is essential to plant breeding programs and therefore for the security of the food supply. At any given time, however, breeders make very little use of this material. Numerous barriers such as temperature, day length, humidity, soil type, and altitude make transfer difficult at best. Hence, even as the number of accessions in germplasm collections has risen over the last decades, use of collections has not kept pace. Collections are often difficult for breeders to use because of poor documentation or evaluation (Marshall 1989).

A central theme in the debate is the limits of private property. When the Undertaking was first initiated, Kloppenburg and Kleinman (1988:188) could well argue that "each side in the debate wants to define the other side's possessions as common heritage." Now, by contrast, each side wishes to define its own possessions as private property. This is relatively easy for those cultivars and plant parts that meet the criteria of utility patents, plant patents, or PVP; however, it has proven much more difficult for the landraces created over millennia by farmers. Several participants in the debate have argued strongly for making landraces private property in the same sense as cultivars produced through scientific breeding (e.g., Kloppenburg and Kleinman 1988; cf. Sedjo 1988). In other contexts, these persons and others have argued that the issue is one of national sovereignty over germplasm as a form of natural resource. It should be emphasized that national sovereignty does not imply *private* property.

Moreover, in all societies property rights are restricted to certain categories. For example, in all nations the sale of body parts is illegal. Most of us still feel uncomfortable with the idea of patenting an animal, though owning one is generally unproblematic. Moreover, certain things are not for sale at any price: the Statue of Liberty and the Louvre, for example. And as Appadurai (1986) has noted, objects take on the status of a commodity only at certain times and in certain contexts. At other times and in other contexts, things are considered sacred and may not be sold or exchanged. Thus we must ask, When and under what conditions, if at all, do we wish to consider germplasm as property? If we do wish to consider it property, should it be private property?

A key issue often ignored is the unequal ability to use germplasm for the public good. Only a handful of Third World nations (e.g., India, Brazil, Mexico) have sufficiently large research systems to use those materials effectively. The obverse of this is that few such nations have the trained scientists needed to create new varieties; hence stronger intellectual property laws are hardly likely to spur indigenous invention.

The issues of germplasm conservation transcend the scientific community; they are issues that ought to be of concern to everyone on the planet. Frankel (1988) is certainly right in noting that during the early years activist groups were nowhere to be found. Yet their lack of concern is traceable, at least in part, to the failure of most scientists to recognize the dimensions of the issue and bring it to world attention. Ironically, though they are detested by many members of the scientific community, people such as Jeremy Rifkin and Pat Roy Mooney have done that community a service. By focusing popular attention on the issue – by making it "political" – they have also helped to increase markedly the financial support that is being provided for it. The issue cannot and should not remain within the scientific community; it is too important.

Finally, we come to the complex issue of farmers' and plant breeders' rights. As Frankel (1988:32) notes: "Landraces are rarely attributable to an individual or even to a particular time or specific location; the term signifies their communal or regional origin, in which natural selection is likely to have played a considerable part." Similarly, today many new cultivars are the product of a corporate endeavor and are not truly traceable to a single breeder. Thus in the case of PBR we are talking about the rights of a corporate entity, while in the case of farmers' rights we are concerned with farmers as a class. Neither position speaks for or against the notion of compensation. That is, compensation for the tasks undertaken may be provided in the form of payment either to a corporate entity or to a class. Similarly, there is no reason why payment to nations in the form of help in conserving germplasm and building effective plant breeding programs cannot be carried out. Farmers' rights are meaningful only if they result in increasing grass-roots capacity to improve plant materials and conserve germplasm. If farmers' rights are interpreted solely in terms of building formal scientific capacity, the concept will further erode farmers' abilities and intensify the crisis (Genetic Resources Action International 1992).

Furthermore, in modern legal systems, which enshrine the concept of rights, we normally talk of the rights of individuals. Rights may include possession or control, use, management, receipt of income, permission to consume, destroy, modify, transfer, distribute, or exclude (Office of Technology Assessment 1986). We may well wish to grant certain of these rights while withholding others.

Moreover, rights are usually associated with responsibilities, which might well include the creation of better varieties as well as the maintenance of germplasm in the field. So we may ask, Is the rights approach the most useful one for germplasm conservation? If so, what are the responsibilities that accompany those rights? Will they be carried out?

11

Conclusion: Toward a Culture of Care

> A litigious world community insisting on sovereign rights to what evolved long before the beginnings of civilization is likely to lose in the long run what it tries to exploit in the short run.
>
> Sir Otto Frankel

This study has taken us far afield in an attempt to depict and understand the historical roots of and contemporary philosophies and strategies for conserving plant germplasm. We have seen that, from the earliest selection of plants, to deliberate breeding for modern agriculture, to the recent use of the methods of biotechnology to create patentable varieties, people have long been intimately involved in making nature. There has been increasing deliberateness in the ways in which individuals, research institutions, corporations, national governments, and international organizations have come to view plant germplasm. This deliberateness has extended from efforts to scientifically collect, store, and manipulate germplasm for ever more specific ends to attempts to control economically and to protect legally and politically rights to germplasm as well.

The social significance of plant germplasm is apparent because it is a product of the work of human beings. The philosophical importance of this fact appears to have escaped or to have been misinterpreted by the courts and patent offices in permitting the legal protection and patenting of these "inventions." Indeed, self-consciously recognizing that human culture began with the conceptual and practical appropriation of plants and seeds for human purposes can lead us to a deeper understanding of contemporary culture: how we conceptually and practically appropriate nature tells us something important about who we are. Western cultures in particular, but most other cultures as well, are at least in part "controllers" of nature. The increasing sophistication with which we control nature is also in part the control of other people and other cultures. Germplasm is simultaneously a biological resource and a social tool. Reflection on this fact should give us pause to consider to what ends that tool is put.

The social significance of germplasm is also revealed in the different ways in which nations regard their natural, social, and scientific resources – and in whether they regard these as resources at all. Indeed, differences in the approach to germplasm collection, maintenance, use, and protection in France, Brazil, Chile, and the United States are a reflection of differences in cultural, political, and even religious backgrounds (table 11.1). Most important, they reflect historical differences in the distribution of economic and social power, both domestically and internationally. It is not surprising that Brazil, with a long history of colonial exploitation, should differ with the United States over intellectual property rights in plant germplasm. Nor is it surprising that France, with its long history of *agriculture paysanne* and *haute cuisine,* should approach traditional landraces in a manner quite different from the United States, the first nation to industrialize agriculture and food through monocultural production and fast-food restaurants. Reflection on human cultural diversity as well as political and economic history should highlight the fact that no one nation's or culture's perceptions of and policies concerning nature, including plant germplasm, are automatically or necessarily the most valid or scientifically, legally, or ethically the best.

Ultimately, the social significance of germplasm is that we, as individuals and as participants in or even observers of organizations and institutions, are continuously making nature. By selecting, collecting, maintaining, and using germplasm, as well as by creating, distributing, marketing, using, and protecting particular cultivars, we are engaging in practices that have had and will continue to have profound consequences that extend far beyond agriculture. Nature itself, including but not limited to the seeds and plants we are naturing, is obviously affected. More important, however, people are affected. Present, future – and perhaps even past – generations of human beings and human cultures have a stake even in the simple action of removing a seed from its parent plant. As reflective individuals have long known, even simple actions can be ethically complex. Indeed, given the array of goals and choices, actions, policies, and

Table 11.1 Germplasm Conservation in Four Nations

	United States	*France*	*Brazil*	*Chile*
Location in world system	World power	World power	At border between more and less developed	Periphery
Type of agriculture	Industrial	Industrial, with remnants of peasant agriculture	Plantation, industrial, and subsistence	Industrial, subsistence
Status of crop germplasm	Most of foreign origin; not linked to biodiversity issues; recently linked to biotechnology	Most domestic; linked to biodiversity issues; linkage to biotechnology being developed	Most domestic, center of diversity; linked to biodiversity and biotechnology	Most introduced; little connection to biodiversity or biotechnology
Official approach to germplasm conservation	Large central seedbanks with few ties to breeders	Decentralized collections linked to breeders	Large central seedbank linked to biotechnology unit	In disarray but headed toward centralization
Role of NGOs	Substantial but not linked to official system	Substantial and integrated into official system	Growing but not linked into official system	Weak and not linked into official system
Role of private sector	Strong for certain crops, not linked to official system	Strong for major crops and integrated into official system	Growing with advent of biotechnology	Generally weak
Position on FAO Undertaking	Party to discussion but has not signed Undertaking	Rapidly became a supporter	Supporter but now with reservations	Supporter
Patent laws with respect to plants and animals	Extended to plants and animals by judiciary	Pressure building to extend legislatively to plants and animals	Being pressured by the United States to extend plants and animals	Plants and animals not patentable
Plant variety protection	Party to UPOV	Party to UPOV	Not recognized	Not recognized
State of food system	Highly industrialized and uniform	Rapidly industrializing	Industrializing rapidly for middle class	Industrialization beginning
State of the environment	Losing diversity rapidly; strong environmental movement	Losing diversity rapidly; strong environmental movement	Losing diversity in some areas; environmental movement growing	Losing diversity in some areas; environmental awareness beginning
Values linked to germplasm conservation	Utility, security of food supply	Utility, preservation of national heritage	Engine of high technology development when linked to biotechnology	Growing concern, especially for export commodities

institutional arrangements associated with it, to unravel the ethical dimensions of germplasm presents a difficult task. Yet it is imperative that we begin to outline some of these ethical issues and responsibilities.

From the Mundane to the Sacred

One of the problems in addressing ethics and ethical responsibilities is that modern science tends to deny or ignore their existence. This may be attributable in part to a desire to "get the job done" or to a lack of appreciation of the ethical dimension of even mundane practices such as collection or cataloging. More likely, however, is that both the current institutional structure of science and the dominant (if outmoded) philosophy of science underlying the plant sciences preclude explicit consideration of ethics (cf. Fuller 1988; Rouse 1987). Specialization and bureaucratization undoubtedly also contribute to a situation in which few individuals ever obtain a "big picture" of the social role of germplasm collection, maintenance, and use.

The failure to appreciate the importance of consideration of the social or ethical aspects of science is, of course, a fundamental tenet of a particular philosophy of science. Many working scientists continue to subscribe, at least implicitly, to a positivist philosophy of science, although it has been rejected by philosophers as fundamentally untenable. Positivism is the philosophy that, among other things, holds that science is value-free and that ethical matters cannot be rationally discussed or analyzed (Fuller 1993). From that perspective, because science contains no value orientation other than to pursue knowledge, deliberate articulation of the political or social foundations or functions of particular scientific activities is inherently nonscientific.

Science, and especially agricultural science, has strong roots in ethical philosophy. Both historically and logically, agricultural science is governed by the philosophy of utilitarianism (Burkhardt 1991). Historically, this philosophy has held that things have value only insofar as they "work": both human artifacts and living things – including perhaps human beings themselves – are to be judged worthy to the extent that they can be used for the production of "maximum utility," a sum of benefits over costs. What counts as a benefit or a cost has long been debated; as this philosophy now stands, however, benefits are defined as "satisfied preferences" of individual human beings. Actions, institutions, practices, even whole cultures are judged, therefore, on the basis of whether "the greatest good of the greatest number" is achieved through them. Whatever ethical responsibilities we might have are dictated by whether we can and do bring about this greatest good. This philosophy has formed the philosophical or ethical foundation for modern agriculture and its attendant science and technology. Agriculture and agricultural research are thought justified by the tremendous gains achieved in improving the material well-being of great numbers of people. Germplasm conservation is conceived as a means toward continued agricultural productivity and is therefore also justified on utilitarian grounds. Our only responsibility in this regard is to be efficient in the provision of usable seed for continued agricultural production.

As pervasive as this viewpoint might be, however, there has been growing unease both with the trade-off or cost-benefit analysis of our institutions and practices and with the underlying notion that things have value only insofar as they produce utility. For example, nature preserves or unspoiled wilderness may have no such utility (Sagoff 1988; Sen and Williams 1982). Some have suggested that the utility might lie in the future as a result of some as yet undiscovered benefit (Schmid 1989). But this is ultimately an unsatisfactory position. As Dahlberg (1987:368–69) explains: "The process of 'discovering' what are seen to be 'new' biological resources is really a process of 'commodification,' that is, the transferring and incorporating into formal and utilitarian systems of species that previously were outside human awareness or were drawn upon by indigenous peoples in ways that acknowledged a mutual symbiosis, rather than a one-way exploitation."

It might be argued, of course, that commodification for utilitarian ends is necessary. In a world in which scarcity is pervasive, it arguably makes sense to think largely in terms of costs versus benefits. A value system structured in this way focuses attention on the problem of scarcity. In a world of potential material abundance – created in part by technoscience for utilitarian ends – the dilemma is that pursuing the "greatest good" through the stepped-up appropriation and commodification of nature may lead to the erosion not only of nature itself but of the very culture that provided us with and fostered the idea that there is a "greater good" than simple satisfaction of preferences or freedom from material wants (Sagoff 1988). Continuing to act and think only in a narrow utilitarian perspective, given what we now know about the natural and cultural effects of doing so, is of questionable ethical merit.

In short, other values, or as Sagoff (1988) suggests, other metaphors beyond utility must also be considered. Perhaps through recapturing a primitive sense of the sacred, or the garden, that is part of our Western heritage, we can learn to appreciate, even cherish, these "nonutilitarian goods." These intuitions or images may provide us with a different model or way of viewing our cultural and agricultural practices and nature itself. They may force us to reassess the responsibilities we have beyond the production of utilitarian goods. No longer may we view our actions and institutions solely on the basis of their productivity. Instead, they must be understood in terms of deeper responsibilities we have to present and future generations and perhaps to nature itself (Gottfried 1992).

Hans Jonas (1985) has argued that the two major approaches to ethics in our time fail to account adequately for the growing dimensions of our responsibilities to the future. Utilitarian cost-benefit analysis and so-called rights theory, he maintains, are conceptually as well as practically unable to give the future its due. No longer is talk of satisfying preferences or of respecting the rights of people in the future enough. Indeed, as critics have long noted, cost-benefit analysis *requires* that future effects of our present actions be "discounted" (Schmid 1989). And many philosophers have noted the conceptual hollowness of the notion that nonexistent future persons can have rights (e.g., DeGeorge 1979). To have a right is to be able to make a claim against others. Those in the future can express neither their preferences nor their indignation. Although we may create in our laws and customs "rights-like" protections for future people, and even for nature itself, we do so on some nonutilitarian and non-rights-based principle (Burkhardt 1990).

Jonas proposed a "new imperative" for the ethical guidance of modern culture: the imperative of responsibility. This is the notion that we must be responsible for the future simply because we create that future. At one level, it is difficult not to accept this general responsibility. Yet how to take it to heart and deliberate and act on it is less than clear. We suggest, however, that his imperative should not be understood in terms of specific obligations we might have to deliberate or to act in particular ways or rules we must follow. Instead, the imperative of responsibility is better understood as an attitude or outlook: in general, but specifically with respect to nature, there is a moral, ethical imperative to care.

"Care," according to standard dictionary definitions, has a variety of meanings: on the one hand, grief, distress, anxiety, concern; on the other hand, attention, heed, caution, watchful oversight, circumspection. An ethic of care includes all these elements. To be sure, attention and circumspection with respect to present and future human beings may conflict neither with maximizing their utility nor with respecting their rights. Grief and concern may lead to a felt responsibility not to violate rights or undermine the environmental quality requisite for the satisfaction of preferences for material comforts. Nevertheless, an imperative to care would redirect our concerns to the full range of consequences that our technoscientific practices impose on present and future generations. It will force us to reassess the relevance of economic welfare or rights in our ethical scheme. A resurrected care for others' lives may lead us to rediscover at least a small part of the sacredness of nature and of our own cultures.

With respect to germplasm, care requires both conservation and preservation, both management and hands-off appreciation. The distinction between conservation and preservation is more a distinction in strategy than one in philosophy (Norton 1987). In reality, both are based on a recognition of our responsibility for and to the ecosystem of which we are a part. We are responsible, in John Locke's (1955 [1690]) words, to use only what we truly need and leave "as much and as good" for posterity: some wild, some compiled. We alone have the capacities for annihilating, debilitating, or enriching whole orders of living things, including ourselves. This power carries the imperative of care.

That we have a responsibility to care for ourselves, the future, and nature is certainly not a given. Sophisticated arguments and considerable reflection are required before the intuitively appealing dimensions of this notion are given more rational, scientific foundation. One such argument runs thus: Let us set aside our particular preferences and biases and ask ourselves what sort of environment and cultural practices, including technoscience, we would think justifiable if we did not know who we were, where, or in what time period we lived. Would it not be reasonable to wish for a culture wherein there were many opportunities for people to experience the richness of life? Would it not be reasonable to wish for an environment populated with an ever-growing diversity of plants and animals and people? To walk, so to speak, in everyman's shoes is to understand the foundations of an ethic of care (Rawls 1971).

Thoughtful utilitarians and rights-oriented ethicists also agree that diversity is generally good. Yet maintaining and promoting diversity can have its drawbacks. For example, George (1988) has argued that in some situations biodiversity is not an unqualified good; in some cases it is precisely the controlled reduction in diversity in the field that brings about the best result for humans. Agriculture is a case in point. Some have suggested the eradication of the boll weevil. Clearly, it is to our advantage that our cotton crop not be ravaged by this plant pest. This can be avoided by employing measures that vastly reduce the population of insects in the field. Yet to bring about the extinction of the species would seem to violate our general ethical responsibility to preserve diversity. Indeed, there are many clear cases of the need for careful reduction in diversity, while the imperative suggests circumspective use and caution in taking these measures too far.

Diversity, of course, can be maintained at many levels. What and which diversity should we conserve? In some cases, our financial resources for conservation or preservation may meet competing demands: landraces or wild relatives? individual mutants or clearly useful cultivars? We cannot hope to resolve these issues here. Let us instead reaffirm the principle that the careful conservation of diversity requires a longer-term vision than the present-focused utilitarian philosophy will admit. Perhaps, in the end, we can even avoid the question of "how much and how good" we can leave the future. There will simply be enough and as needed.

The question of what we should conserve is inextricably bound to the problem of how to conserve. If care is to be the central concern in germplasm conservation, and if our obligation is to species rather than to individuals, then our responsibility must extend to everything in nature. No longer can priorities be established solely on the current value of an organism to us. Yet we can establish short- and long-term priorities. Short-term priorities can be based on the means we have available immediately and the urgency of the situation (i.e., the likelihood of extinction if we do not act). This further suggests that no one means will be the sole solution for all organisms in danger of extinction. Nor can any single institution or single nation take on the task itself. Proponents of germplasm banks must begin to recognize that much material can be preserved only in its natural habitat. Conversely, proponents of nature preserves must recognize that the evolution of certain species is so closely linked to human culture that once we cease cultivating them, we can only preserve those materials in germplasm banks. Similarly, proponents of state control must recognize the fallibility and temporality of states, while proponents of voluntarism must recognize the importance and value of state intervention. In the longer term, however, neither of the two solutions is likely to solve the problem. As long as our culture is pitted against nature rather than a part of it, no amount of financial resources and scientific expertise is likely to reverse the trend toward reduced diversity. The long term will require a thorough rethinking of the goals and direction of human civilization, an admittedly enormous task.

Our responsibilities do not end, however, when we decide what germplasm we shall appropriate from the wild. For even the most routine activities such as cataloging, evaluation, storage, and the like begin to carry more ethical weight. Conservation ex situ is justified only if due care is also exercised in ensuring the integrity of accessions, the fairness and completeness of descriptor sets, and the adequacy of storage facilities. If we have judged that conservation rather than preservation, for a particular species or subspecies, is the only way to appropriate nature for our careful purposes, we must make certain that those purposes are indeed served in an efficient – and just – manner. In the final analysis, for each of our activities regarding germplasm, we must ask ourselves, How can our collective, institutional care and concern for the world of nature be revitalized? How can we take on the imperative to care?

Increasing Care and Concern

Until recently it appeared that germplasm conservation had little or no utility – the central measure of worth – and there was little incentive to finance it. This situation must be reversed. In the past, agricultural research institutions have been extraordinarily effective in building the clienteles necessary to support important research endeavors. For example, the U.S. land grant universities actively organized farmers during the second and third decades of this century to support agricultural research and education (Busch and Lacy 1983). Even today the importance of client groups to the effective functioning of agricultural research systems throughout the world cannot be underestimated. Though some have rightly criticized the narrowness of the constituencies that these groups often represent (Hightower 1973), they are not a luxury to be discarded; they are instead an essential part of the effectiveness of agricultural scientific institutions. The proposals we offer are modest ones that might be articulated before legislative and administrative bodies so as to ensure that we do not destroy diversity and that we live up to our responsibilities to the future and to nature itself.

FOOD SECURITY

Though the nations of the world spend billions of dollars each year on weapons that will supposedly protect us against each other, we spend little on germplasm conservation. Plucknett et al. (1987) estimated that worldwide germplasm expenditures annually total only $55 million. A single submarine, by comparison, now costs considerably more. Given that the continued production of our crops and livestock depend upon the ability to respond rapidly to diseases and pests, and that this in turn depends upon our ability to maintain a diverse stock of germplasm, it is obvious that germplasm conservation must receive higher priority on grounds of national defense. Trading one submarine for better germplasm maintenance would be a bargain indeed.

BUILDING NEW CONSTITUENCIES FOR GERMPLASM MAINTENANCE

Six major groups have a common interest in germplasm conservation: nation-states, seed and livestock breeding companies and their parent corporations, voluntary associations of seed savers, environmental groups, farmers, and plant and animal breeders. Nations have an interest and have done much to establish germplasm banks and nature preserves over the last several decades. Though there are clear areas of dispute among nations, there are far more points of agreement. Yet few nations have drawn on the potential support that exists or may be created within the other five groups. For example, seed and livestock companies have a vested interest in ensuring that adequate stocks of germplasm are available for use in future breeding projects. They can be strong supporters of government financing of germplasm conservation, especially for germplasm banks.

In many nations there also exist voluntary associations of seed savers and others concerned about germplasm. Currently, their activities are largely divorced from those of governments. Yet they are a natural ally in conservation. Simple pamphlets on obtaining and growing unusual plant varieties, distributed with government funds, would serve both to increase public awareness and to encourage active public participation in conservation programs. For relatively small sums, such private voluntary associations might become distribution centers for rare varieties. The key ingredients in such a program are public education and conservation. Such a program would encourage duplication and redundancy. It would not involve storage but rather the active planting and growing of unusual varieties. It would serve to build public support while appealing to the desire on the part of many to grow something exotic.

Some scientists might argue that such groups are inconsequential and unlikely to understand the need for special care in maintaining exotic species. Others would argue that their informal status makes them poor keepers of our genetic heritage. The facts appear to belie this position. Many of the members of these groups are already actively consulted in regard to the conservation of some species. In addition, illiterate peasants not only conserved germplasm for centuries but made major improvements. To think that our much better educated population of today is incapable of doing that is to be victim to an erroneous scientific elitism. Moreover, though we like to think of our government institutions as permanent and enduring, we should remember that the record of history shows quite the opposite.

Environmental groups should also be drawn into the debates over germplasm conservation. Some of these groups are already concerned about large animals (e.g., the World Wildlife Fund). Ironically, whales receive enormous attention while seeds are virtually ignored. More education is needed that shows how size can blind us to the many other species that face extinction. Unlike the smaller seed saver groups, these large environmental organizations have already shown that they can mobilize large numbers of persons and significantly influence government policies. Their support for germplasm conservation is vital.

Both seed savers and environmental groups could benefit from stronger linkages with botanical and zoological gardens. With proper funding, gardens would serve a much greater educational and client-building role than they currently do. In particular, gardens in Third World nations could educate wealthy tourists (in return for a contribution) about the gravity of the problem of species loss.

Farmers are yet another group with a potentially strong interest in germplasm conservation. Indeed, farmers have much to lose immediately if germplasm is not conserved. Yet the importance of germplasm conservation will have to be explained to farmers because most are unaware of its necessity even when they actively participate in it through the cultivation of landraces.

Finally, plant and animal breeders are a logical constituency for germplasm conservation. Most, however, are likely to become ardent supporters only to the extent that germplasm is not merely conserved but evaluated as well. This suggests that new ways need to be developed to engage breeders actively in the evaluation process. The French approach (chapter 7) bears study by other nations. And appeals to breeders should clearly emphasize the inadequate support for evaluation as well as pointing to the successes with a few crops that have been well studied (e.g., tomatoes).

TAXING SEEDS AND ANIMALS

A third possibility is a tax, the proceeds from which would be specifically devoted to germplasm conservation. Such a tax could be levied on the sales of bulk seed, semen, and breeding stock. Given the relatively low cost of germplasm conservation as a proportion of seed and animal sales – one estimate suggests that $100 million could be raised by a tax equal to 0.3 percent of seed sales (Barton and Christensen 1988) – such a tax would be virtually unnoticeable to the purchasers or the general public. Any country could begin immediately to impose such a tax with the goal of having a worldwide tax by some target date (for example, 10 years). Proceeds from the tax would be used to support collection, creation, and maintenance of genebanks and creation of nature preserves. Some system also might be devised to help more Third World nations to establish their own institutions for the conservation of genetic resources. Revenues from such a system would grow in proportion to the volume of improved seeds and animals produced. Thus support for conservation would increase in proportion to the replacement of landraces by higher-yielding organisms.

RELINKING FOOD, AGRICULTURE, AND THE ENVIRONMENT

For the last 300 years we have sought to separate food, agriculture, and the environment. We have worked hard to make food a mysterious thing purchased in the supermarket, to make nature into the environment – something utterly external to us. We now need to move in the other direction. We need to develop new ways of showing consumers the ways in which food gets to them and the agricultural and environmental choices made in food consumption. We also need to show environmentalists that working landscapes, and especially agricultural landscapes, are themselves worthy of conservation, that the rural environment *is* largely agriculture. And we need to show farmers that their responsibilities do not end at the farm gate, that they are part of a much larger social system in which what they do affects us all.

REVALUING TRADITIONAL KNOWLEDGE

A misconception often found in the West is that traditional knowledge is mostly superstition, wrong, and irrelevant to the needs of the modern world. Unfortunately, even the leaders of many non-Western nations share this view, having themselves been trained in the West and taught to share this belief (Goonatilake 1982). One tragic result of what Gadamer (1975) in another context has called the modern "prejudice against prejudice" is the destruction of the stock of knowledge of indigenous plants, animals, and soils in many parts of the world. Yet this knowledge can provide us with many important insights into the natural world developed over centuries, even millennia, by trial and error and observation. Admittedly, this knowledge does not fit easily into modern taxonomic schemas. It is often very specific to particular ecological niches. Nevertheless, it is far more developed than any modern taxonomy for the realms of nature that it maps. Moreover, it often incorporates an understanding of nature as a whole, of which human beings are but a part. By codifying and synthesizing indigenous knowledge into our Western worldview, we perhaps can come to understand the natural world better, educate ourselves about the infinite diversity that surrounds us, and even learn more about how diversity may be conserved. In short, the conservation of natural diversity requires the conservation of cultural diversity.

EDUCATING THE PUBLIC ABOUT BIODIVERSITY

Currently, it is an understatement to say that the public in industrialized nations is uninformed about biodiversity issues. Most residents of industrialized countries and many of the elites of developing nations are now divorced from the process of food production. Even those of us who proclaim our love for nature and enjoy an occasional hike in the woods or swim in a mountain stream are quite content to return to our urban dwellings and the comforts of home. Hence both agriculture and nature have become abstractions – perhaps worthy of preservation but unconnected to our daily concerns.

This suggests that public education cannot merely involve the use of the media. Public education must be made immediate; it must involve the public directly in the conservation process by demonstrating its importance and by making opportunities available for direct participation. Several options are already available.

Written Guides for Amateur Germplasm Collectors. The excellent guide by Marchenay (1987) is a model. It provides suggestions for the amateur on how to locate exotic materials, rudimentary guidelines for classification, and directions on where to send materials for further classification and storage. Similar guidebooks in other languages are desperately needed.

Historical Parks. Virtually every industrialized nation and many developing nations maintain historical parks where the homes, farms, and crafts of yesteryear are collected and often displayed in action. A few have incorporated living collections of heirloom crops into their exhibits and demonstrations. With a little encouragement, many more could do the same thing.

Elementary and Secondary School Projects. Many of the current biological science projects in elementary and secondary schools are little more than rote repetitions of classical experiments. Moreover, they are divorced from context and do little to stir the imagination of students. Simple experiments illustrating principles of plant growth and natural selection would help students to build a greater appreciation of the importance of both agriculture and genetic diversity. Farm organizations would be the logical groups to take on the development of such project kits for distribution to public school biology and social studies teachers. Doubtless, other activities that also provide opportunities for participation could be easily devised.

SCIENCE AND DEMOCRACY

The call for participation in the conservation of biodiversity calls forward yet another underlying issue, one too large to treat adequately here: the need to democratize science. Throughout this volume we have argued that technoscience involves choices – choices about what aspects of the natural world shall be revealed as well as choices about who we shall be. The Baconian ideal of science as a substitute for politics is fundamentally misguided. As Latour (1987) has suggested, science is politics by other means. As such it has no special claim to exemption from democratic norms and values. The conservation of biodiversity – and consequently of human cultural diversity – is too large and too complex a task to delegate to a small cadre of scientists. It demands the opening of scientific decision making to the broader public.

This is not to suggest that scientific decisions can be made by popular vote. Such a vulgar notion of participation is both absurd and misleading. Equally misleading, however, is the notion that scientists must be left to make such decisions on their own. Scientists have no special status that would permit them to make more valid decisions as to what must be done.

THE PATENT DEBATE

Perhaps the most difficult point on which to make a recommendation here concerns patents. Yet we must address the issues. We may begin with the following propositions: Most of the scientific work that makes it possible to have this debate was accomplished using public funds. Without the huge investments of the U.S. National Institutes of Health and other agencies in the United States and other nations, there would be no squabble over patents. In a capitalist world, there is general agreement that companies ought to be able to earn a fair return on their investments. The patent system was designed *not* to ensure that companies make a profit but to serve the public good by encouraging invention and innovation. Intellectual property rights are not natural rights but rights granted by statute. The patent system was not designed with life forms in mind.

These propositions together suggest the following:

- Life forms are sufficiently different from other subject matter for patents that extension of intellectual property legislation to cover them should be a matter for legislation, not the courts.
- Intellectual property rights, if they are to be extended to life forms, should be of such length and scope as to permit the recovery of investment plus a reasonable return on that investment. This procedure is analogous to that currently used to regulate public utilities (Office of Technology Assessment, 1986).
- Licensing of biotechnological inventions should be mandatory so as to permit new entrants into the technical arena and to prohibit lengthy monopolies over particular cultivars.

SCIENCE AND THE THIRD WORLD

Most Third World nations have no chance of becoming significant actors in producing biological innovations unless they receive considerable help from the developed nations to build adequate scientific communities. Indeed, as Gaillard (1991) has shown, even the best of Third World scientists are short of the minimum needed to participate in the world scientific community as more than spectators. Curiously, millions of dollars have been spent to train scientists from developing nations in the United States and Europe, but there is considerable reluctance to provide annual support even in cases where it is unrealistic for national governments, already operating under heavy debt burdens, to pay the bill. Even the international agricultural research centers have a budget smaller than that of most American states.

Though there are compelling moral reasons for doing it, including our responsibility for the other two-thirds of humankind, we are not optimistic about the willingness of developed nations to finance agricultural research in the Third World. Indeed, it can even be argued that agricultural research in the Third World would be an excellent long-term investment, raising living standards and the ability of the Third World to engage in trade with the West. Yet at the moment the West appears to be caught in its own immediate economic problems and is likely to see agricultural research in the Third World as a very low priority.

We must take seriously the words of one Brazilian scientist who argued: "It doesn't make a big difference who coordinates the international network of genebanks; what counts is who has more capacity to manipulate and use plant genetic information. However, support for conservation without helping developing countries to increase their scientific and technological capacity to use their plant genetic resources is a crime." His words ring true.

NEW MODELS OF DEVELOPMENT

Beyond these specific remedies, we must begin to question the model of development that we in the West have embraced these past 300 years and have foisted upon many of the peoples of the rest of the world. We refer here not to a specifically economic model such as capitalism or socialism but to the underlying philosophical assumptions upon which Western development – both capitalist and socialist – is based. The former Soviet republics are no better than we are despite their adherence until recently to a different ideological position. Thus what we are talking about here is far deeper than any mere ideology. It is a set of root practices, beliefs, and myths about the abilities that we possess to restructure the entire planet.

Such a model of development must operate at the international, national, and local levels simultaneously. It must provide for international cooperation and competition. It must recognize the different socioeconomic locations of each nation. And it must grant to local communities both the rights and responsibilities for the maintenance and use of germplasm. Perhaps most important, it must take into account that "the loss of biodiversity will become a global crisis for the rural poor. For these people, the loss of biodiversity translates into loss of food, construction materials, medicine, fuel, and material inputs essential to their survival" (Alcorn 1991:331). Creation of new models will take all of the best minds that the world has. And we need to get on with the task immediately.

Manufacturing Plants

We are now on the verge of yet another set of major changes in the nature of what we grow, what we eat, and consequently who we are. Unlike the previous changes described above that were introduced without much thought as to their consequences, a set of choices is open to us. The new biotechnologies can be used to manufacture plants in manufacturing plants (Rogoff and Rawlins 1987). Or they can be used to reunite food and agriculture in a new way. Let us state what we are not proposing first: there is no way that biotechnology can provide us with a technical fix, a simple way out of the current dilemmas that confront us. Nor can we say that biotechnology represents technology out of control; technology is out of control only if a factory explodes or a vat leaks. Otherwise, technology is always under the control of someone or some organization. Yet the new biotechnologies do reinforce the urgency of the question, What kind of nature do we want?[1] If we were to answer that question collectively, then we could ask what biotechnologies might be useful in helping us to achieve that kind of nature. We may go even further still: because the way we treat nature is indicative – no, an essential part – of the way we treat each other, the nature we want must be one that is humane, caring, and befitting of ourselves as moral beings. This, we submit, can be accomplished only by reuniting food and agriculture once more. Moreover, this cannot be done through some mass return to the land. Industrialization, urbanization, and population growth have already eliminated that possibility. It will require instead that we develop new institutional mechanisms to link food and agriculture, institutions that allow us to show our care for each other through our reverence for nature. The need for these institutions is manifested every time someone looks into a petri dish and sees a new form of culture. The form that culture takes reveals something about both the cells in the dish and us, for in the final analysis there is no way to separate its cultural evolution from ours.

No matter how well we may run our germplasm banks, no matter how many protected wilderness areas we may create, no matter how many species we protect, we shall never be able to take on the active responsibility for them all. That is simply beyond the capabilities of the human species. Perhaps we should abandon our war with nature and reconsider the limits of human ingenuity and – if it is not already too late – the indigenous knowledge that the non-Western peoples of the world have. This is a challenge of perhaps greater magnitude than any mere system for conserving genetic diversity. Yet it is essential that we respond to it if we are to take our responsibilities to other species and to our own future generations seriously. It is essential if our culture is to care.

Notes

1. Constituting Nature, Manufacturing Plants

1. The term "plant germplasm" is commonly used by scientists to refer to the full range of genetic materials that are used by people to reproduce plants. Thus plant germplasm is largely seeds but also includes roots, cuttings, and other plant parts from which whole plants may be reproduced.

2. Throughout this volume we make use of Latour's neologism, "technoscience," to emphasize the difficulty one has in separating science and technology in contemporary settings.

2. Properties of Culture

1. As Latour (1993) notes, all knowledge is applied locally, including that labeled "universal." By "local" we mean that knowledge not easily transferable to other sites.

2. The rediscovery of Mendel suggests that Mendel knew what he discovered. In fact, the evidence suggests that his findings make sense only in light of a later debate (Brannigan 1981; Monaghan and Corcos 1990).

3. Stages are social constructions; time flows continuously.

4. The term *genebank* is somewhat misleading because what is being stored is seeds or other plant parts, not individual genes.

5. Chang (1989) has noted that larger collections are more likely to contain the traits desired by breeders. Large collections also are likely to cost less, are more visible, and are more likely to obtain public support.

6. Chatelin (1979) makes much the same point for soil classification systems.

3. A Competition-Cooperation Paradox

1. The Keystone International Dialogue Series on Plant Genetic Resources played an important role in changing the tone of the debate at FAO and later at UNCED. The Keystone Center is a nonprofit organization which facilitates the resolution of public policy

conflicts through the use of a consensus dialogue approach. The dialogue series involved participants from nongovernmental organizations, national governmental organizations, corporations, research institutions, and international and intergovernmental organizations from developing countries (Keystone Center 1991). The result was a global initiative that called for the establishment of a nonprofit organization to fund and support research to preserve plant germplasm.

2. Intellectual property rights include patents, copyrights, trademarks, and trade secrets in addition to more specialized forms of protection for plants. See Chapter 10.

6. The View from the Grass Roots

1. The questions raised in this dispute are similar to those raised by art critics. The authenticity of a work of art and of an heirloom seed variety is of considerable import to those who preserve these aspects of our cultural heritage.

7. France

1. The Common Catalog is merely the sum of all of the catalogs of the various members of the European Community.

2. In principle, even a variety discovered and left unchanged by the breeder may be granted a certificate if it meets the DHS criteria. In practice, however, this would be a highly unusual case.

3. Matters are confused somewhat because the European Patent Convention members are not exactly the same states that belong to the Community. Because France belongs to both, this difference need not concern us here.

4. Forest trees and vegetables are exempt from tests of cultural value. Fruit trees must undergo this test only for renewals after the first ten years in the catalog.

5. The advisability of grouping open-pollinated landraces into population pools as a conservation strategy has been strongly criticized by certain population biologists within INRA (e.g., Olivieri and Prosperi 1989). The issue is too complex to discuss fully here.

6. The Association maintains over 250 species of fruits (mainly apples) (SOLAGRAL 1988). When one considers that 93 percent of French apple production today is of American origin and 71 percent is Golden Delicious (Grall and Levy 1985), the importance of such a project is self-evident.

8. Brazil

1. There is some debate as to whether the British violated Brazilian law by taking the seeds out of the country. Brockway (1979) argues that the seeds were smuggled out. Smith (1985) and Plucknett et al. (1987) argue that the law in question had not yet been passed when the British removed the seeds.

10. Cultures of Property

1. The Strasbourg Convention of 1963 distinguished between "microbiological" products, which were patentable, and "essentially biological" products, which were not. The European Patent Convention of 1973 specifically excluded protection for plants and animals, thereby reinforcing the division of labor with UPOV. The United States did not follow suit.

2. Pioneer Hi-Bred, the company that employed both authors at the time they wrote the article, has been among the very few to examine seriously and participate in Third World markets.

3. Mody (1990) has argued that this figure is probably inflated. It is countered by a World Bank estimate that current trade protection by the industrial nations costs the developing world $50 billion in lost export revenues (Pinstrup-Andersen 1992).

11. Toward a Culture of Care

1. This is not to say that we have the capability of deciding *precisely* what kind of nature we want. This, too, would be a naive form of technological utopism. But we can – perhaps must – decide in what general direction to go or risk destroying the nature of which we are a part.

Selected Bibliography

Adair, James. 1975. Indian Agriculture, 1755. In *Agriculture in the United States: A Documentary History*. Ed. Wayne D. Rasmussen, 1:72–75. New York: Random House.

Adelmann, Arllys, Steve Demuth, Becky Idstrom, Joanne Thuente, and Kent Whealy. 1992. *Seed Savers 1992 Harvest Edition*. Decorah IA: Seed Savers Exchange.

Adler, Reid. 1986. Can Patents Coexist with Breeders' Rights? Developments in U.S. and International Biotechnology Law. *International Review of Industrial Property and Copyright Law* 17(2): 195–227.

Agassiz, Louis. 1867. *A Journey in Brazil*. Boston: Houghton Mifflin.

Agricultural Research Service. 1990. *Seeds for Our Future: The U.S. National Plant Germplasm System*. Beltsville MD: USDA.

——— 1992. *U.S. National Genetic Resources Program. Initial Report*. Washington DC: USDA.

Aguiar, Ronald Conde. 1986. *Abrindo o pacote technólogico: Estado e pesquisa agropecuária no Brasil*. São Paulo: Polis/CNPq.

Akihama, T., and K. Nakajima. 1978. *Long Term Preservation of Favourable Germ Plasm in Arboreal Crops*. Tokyo: Fruit Tree Research Station, Ministry of Agriculture and Forestry.

Albuquerque, Rui H., Antonio C. Ortega, and Baastian P. Reydon. 1986a. O setor público de pesquisa agrícola no Estado de São Paulo; parte I. *Cadernos de Difusão de Tecnologia,* 3(2): 79–132.

——— 1986b. O setor público de pesquisa agrícola no Estado de São Paulo; parte II. *Cadernos de Difusão de Tecnologia,* 3(2): 243–96.

Alcorn, Janis B. 1991. Ethics, Economies, and Conservation. In *Biodiversity: Culture, Conservation, and Ecodevelopment*. Ed. Margery L. Oldfield and Janis B. Alcorn, pp. 317–49. Boulder CO: Westview.

American Farmer. 1975. Beginnings of the Farm Press. In *Agriculture in the United States: A Documentary History*. Ed. Wayne D. Rasmussen, 1:446–47. New York: Random House.

Appadurai, Arjun. 1986. Introduction: Commodities and the Politics of Value. In *The Social Life of Things: Commodities in Social Perspective*. Ed. Arjun Appadurai, pp. 3–63. Cambridge: Cambridge University Press.

Arbez, Michel, ed. 1987. *Les ressources génétiques forestières en France*. Vol. 1: *Les conifères*. Paris: INRA/BRG.

Arnal, Yolanda Texeira. 1987. Explorados botanicos Europeus en Venezuela durante el siglo XIX. *Quipu* 4(2): 185–211.

Ashton, Peter S. 1988. Conservation of Biological Diversity in Botanical Gardens. In *Biodiversity*. Ed. E. O. Wilson, pp. 269–78. Washington DC: National Academy Press.

Auerbach, Bradford C. 1983. Biotechnology Patent Law Developments in Great Britain and the United States: Analysis of a Hypothetical Patent Claim for a Synthesized Virus. *Boston College International and Comparative Law Review* 6 (spring): 563–90.

Bacon, Francis. 1974. *The Advancement of Learning and the New Atlantis*. Oxford: Clarendon Press.

Bajaj, Y. P. S. 1981. Regeneration of Plants from Potato Meristems Freeze-Preserved for 24 Months. *Euphytica* 30:141–45.

——— 1984. The Regeneration of Plants from Frozen Pollen Embryos and Zygotic Embryos of Wheat and Rice. *Theoretical and Applied Genetics* 67:525–28.

Bannerot, Hubert. 1986. L'evolution de l'amélioration des variétés de légumes. In *La diversité des plantes légumières: Hier, aujourd'hui et demain*. Ed. Bureau des Ressources Génétiques, pp. 53–64. Paris: Bureau des Ressources Génétiques.

Bannerot, Hubert, and C. Foury. 1986. Utilisation des ressources génétiques et création variétale. *Bulletin Technique d'Information* (Ministry of Agriculture) 407:93–105.

Bartélemy, Guy. 1979. *Les jardiniers du roy: Petite histoire du Jardin des Plantes de Paris*. Paris: Le Pélican.

Barton, John H. 1991. Patenting Life. *Scientific American* 264 (March): 40–46.

Barton, John H., and Eric Christensen. 1988. Diversity Compensation Systems: Ways to Compensate Developing Nations for Providing Genetic Materials. In *Seeds and Sovereignty*. Ed. Jack R. Kloppenburg Jr., pp. 338–55. Durham: Duke University Press.

Basalla, George. 1967. The Spread of Western Science. Science 156 (3 May): 611–22.

Bathe, Greville, and Dorothy Bathe. 1935. *Oliver Evans: A Chronicle of Early American Engineering*. Philadelphia: Historical Society of Pennsylvania.

Beier, F. K., R. S. Crespi, and J. Straus. 1985. *Biotechnology and Patent Protection: An International Review*. Paris: Organisation for Economic Cooperation and Development.

Bent, Stephen A., Richard L. Schwab, David G. Carlin, and Donald D. Jeffrey. 1987. *Intellectual Property Rights in Biotechnology Worldwide*. New York: Stockton Press.

Berger, John. 1979. *Pig Earth*. New York: Pantheon.

Berlan, Jean Pierre. 1987. Recherches sur l'economie politique d'un changement technique: Les mythes du maïs hybride. Ph.D. dissertation, Les Milles: Université Aix-Marseilles II.

Berlan, Jean Pierre, and Richard Lewontin. 1986a. The Political Economy of Hybrid Corn. *Monthly Review* 38 (July–August): 35–47.

——— 1986b. Breeders' Rights and Patenting Life Forms. *Nature* 322 (28 August): 785–88.

Bidwell, Percy W., and John I. Falconer. 1941. *History of Agriculture in the Northern United States*. New York: Peter Smith.

Bijker, Wiebe, Thomas P. Hughes, and Trevor Pinch, eds. 1987. *The Social Construction of Technological Systems: New Directions in the Sociology and History of Technology*. Cambridge MA: MIT Press.

Biotechnology and Development Monitor. 1991. Brazil to Recognize Biotechnology Patents. *Biotechnology and Development Monitor* 8 (September): 18.

Borges, Jorge Luis. 1962. *Ficciones*. New York: Grove Press.

Boucharenc, Stéphane. 1988. Le Bureau des Ressources Génétiques: Premier objectif, sensilibiser. *Ressources Génétiques et Développement* (SOLAGRAL). 1 (May): 4.

Brannigan, Augustine. 1981. *The Social Basis of Scientific Discoveries*. Cambridge: Cambridge University Press.

Braudel, Fernand. 1973. *Capitalism and Material Life, 1400–1800*. New York: Harper.

Brockway, Lucile H. 1979. *Science and Colonial Expansion: The Role of the British Royal Botanic Gardens*. New York: Academic Press.

Brossier, Jacques. 1986. L'inscription aux catalogues officiels et la protection des obtentions végétales. In *La diversité des plantes légumières: Hier, aujourd'hui et demain*. Ed. Bureau des Ressources Génétiques, pp. 67–73. Paris: Bureau des Ressources Genétiques.

Brown, A. H. D. 1989. The Case for Core Collections. In *The Use of Plant Genetic Resources*. Ed. A. H. D. Brown, O. H. Frankel, D. R. Marshall, and J. T. Williams, pp. 136–56. Cambridge: Cambridge University Press.

Brown, William L. 1988. Plant Genetic Resources: A View from the Seed Industry. In *Seeds and Sovereignty*. Ed. Jack R. Kloppenburg Jr., pp. 218–30. Durham: Duke University Press.

Buret, Philippe, François Boulineau, and Richard Brand. 1986. Maintien et evolution d'une variété du domaine public chez une espèce allogame: Exemple du radis 'de 18 Jours.' In *La diversité des plantes légumières: Hier, aujourd'hui et demain*. Ed.

Bureau des Ressources Génétiques, pp. 95–98. Paris: Bureau des Ressources Génétiques.

Burkhardt, Jeffrey. 1990. The Morality Behind Sustainability. *Journal of Agricultural Ethics*. 2 (spring): 113–28.

——— 1991. The Value Measure in Agricultural Research. In *Beyond the Large Farm*. Ed. Paul Thompson and W. Stout, pp. 79–106. Boulder CO: Westview.

Burkhardt, Jeffrey, L. Busch, and W. B. Lacy. 1988. The Ethics of Germplasm. *Diversity*, no. 13, pp. 25–27.

Busch, Lawrence, and William B. Lacy. 1983. *Science, Agriculture, and the Politics of Research*. Boulder CO: Westview Press.

——— 1984a. Sorghum Research and Human Values. *Agricultural Administration* 15:205–22.

——— 1984b. *Food Security in the United States*. Boulder CO: Westview Press.

——— eds. 1986. *The Agricultural Scientific Enterprise: A System in Transition*. Boulder CO: Westview.

Busch, L., W. B. Lacy, J. Burkhardt, and L. R. Lacy. 1991. *Plants, Power and Profit: Social, Economic, and Ethical Consequences of the New Biotechnologies*. London: Basil Blackwell.

Busch, Lawrence, and Carolyn Sachs. 1981. The Agricultural Sciences and the Modern World System. In *Science and Agricultural Development*. Ed. L. Busch, pp. 131–56. Montclair NJ: Allanheld, Osmun.

Buttel, Frederick. 1987. Some Observations on North-South Issues in Genetic Resources Conservation and Exchange. Staff Paper, Cornell University, Ithaca NY.

Buttel, Frederick H., and Jill Belsky. 1987. Biotechnology, Plant Breeding, and Intellectual Property: Social and Ethical Dimensions. *Science, Technology, and Human Values* 12 (winter): 31–49.

Carleton, Mark Alfred. 1900. *The Basis for the Improvement of American Wheats*. Washington DC: USDA, Division of Vegetable Physiology and Pathology, Bulletin 24.

Carter, Clarence E. 1934. *The Territorial Papers of the United States*. Washington DC: U.S. Government Printing Office.

Casa Valdés, Teresa Ozores y Saavedra, marquesa de. 1987. *Spanish Gardens*. Woodbridge, England: Antique Collectors' Club.

Cauderon, André. 1980. Sur la protection des ressources génétiques en relation avec leur surveillance, leur modelage, et leur utilisation. *Comtes Rendus des Séances de l'Academie d'Agriculture* 66(12): 1051–68.

——— 1981. Sur les approches ecologiques de l'agriculture. *Agronomie* 1(8): 611–16.

——— 1982. Ecological Approaches to Horticulture and Genetic Diversity. *Chronica Horticulturae* 22(3): 46–47.

——— 1983. A Chronicle of Thirty Years of Maize in France: Genetics, Breeding, and Expansion. *Plant Variety Protection* 33 (April): 41–49.

——— 1984. Ressources génétiques, amélioration des plantes et agriculture. *Bulletin Technique d'Information* (Ministry of Agriculture) 391:385–90.

——— 1985. Un projet de Centre Français de Ressources Génétiques pour les Céréales. *Comtes Rendus de l'Académie d'Agriculture de France* 71(8): 809–20.

——— 1986. Voies actuelles pour la diversification génétique des productions légumières. In *La diversité des plantes legumières: Hier, aujourd'hui et demain*. Ed. Bureau des Ressources Génétiques, pp. 213–18. Paris: Bureau des Ressources Génétiques.

——— 1989. Pour une politique de la diversité biologique. Le Monde (14 June): 21.

CENARGEN. 1988. Centro Nacional de Recursos Genéticos e Biotecnologia: Perfil. Unpublished document.

CENARGEN Informa. 1987. CENARGEN obtém gêmeos idênticos de bovinos por micromanipulação. *CENARGEN Informa,* no. 2 (October–December): 6.

Chang, T. T. 1987. The Availability of Crop Germplasm. In *Crop Exploration and Utilization of Genetic Resources,* pp. 225–31. Changhua, Taiwan: Taichung District Agricultural Improvement Station.

——— 1989. The Case for Large Collections. In *The Use of Plant Genetic Resources*. Ed. A. H. D. Brown, O. H. Frankel, D. R. Marshall, and J. T. Williams, pp. 123–35. Cambridge: Cambridge University Press.

Chang, Te-Tzu, Sherl M. Dietz, and Melvin N. Westwood. 1989. Maintenance and Use of Plant Germplasm Collections. In *Biotic Diversity and Germplasm Preservation: Global Imperatives*. Ed. Lloyd Knutson and Allan K. Stoner, pp. 127–59. Dordrecht: Kluwer Academic Publishers.

Chapman, C. G. D. 1989. Collection Strategies for the Wild Relatives of Field Crops. In *The Use of Plant Genetic Resources*. Ed. A. H. D. Brown, O. H. Frankel, D. R. Marshall, and J. T. Williams, pp. 263–79. Cambridge: Cambridge University Press.

Chapman, Jefferson, Hazel R. Delcourt, and Paul A. Delcourt. 1989. Strawberry Fields, Almost Forever. *Natural History* 98(9): 51–59.

Chatelin, Yvon. 1979. *Une epistemologie des sciences du sol*. Paris: ORSTOM, mémoires no. 88.

Chatelin, Yvon, and Gerard Riou, eds. 1986. *Milieux et paysages: Essai sur diverses modalités de connaissance*. Paris: Masson.

CIRF. 1986. Brasil. In *El germoplasma vegetal en el paises del cono Sur de America Latina*. Ed. M. Cerezo-Mesa and Jose T. Esquinas-Alcazar, pp. 67–132. Rome: CIRF (Consejo International de Recursos Fitogeneticos).

Clarke, Adele E., and Joan H. Fujimura. 1992. *The Right Tools for the Job: At Work in Twentieth-Century Life Sciences*. Princeton: Princeton University Press.

Clout, Hugh D. 1983. *The Land of France, 1815–1914*. London: George Allen & Unwin.

Cochrane, Willard. 1979. *The Development of American Agriculture: An Historical Analysis*. Minneapolis: University of Minnesota Press.

Colden, Jane. 1963. Botanic manuscript. Ed. H. W. Rickett and Elizabeth C. Hall. New York: Garden Club of Orange and Dutchess Counties.

Comte, Françoise. 1989. The Position of COPA and COGECA on the Legal Protection of Biotechnology Inventions. In *Patenting Life Forms in Europe*. Ed. International Coalition for Development Action, pp. 38–40. Brussels: International Coalition for Development Action.

Contreras, Andrés. 1987. Mejorimiento de cultivos para pequeños agricoltores. Informe Final prentado al Centro Internacional de Investigaciones para el Desarrollo del Canada. Valdavia, Chile.

Copeland, L. O., and McDonald, M. B. 1985. *Principles of Seed Science and Technology*. 2d ed. New York: Macmillan.

Cornet, René Jules. 1965. *Les phares verts*. Brussels: Editions L. Cuypers.

Council for Agricultural Science and Technology. 1985. *Plant Germplasm Preservation and Utilization in U.S. Agriculture*. Report 106. Ames IA: CAST.

Country Life Commission. 1911 [1909]. *Report of the Commission on Country Life*. New York: Sturgis and Walton.

Crespi, R. Stephen. 1988. *Patents: A Basic Guide to Patenting in Biotechnology*. Cambridge: Cambridge University Press.

Crosby, Alfred W. 1986. *Ecological Imperialism: The Biological Expansion of Europe, 900–1900*. Cambridge: Cambridge University Press.

Crouch, Martha L. 1990. Debating the Responsibilities of Plant Scientists in the Decade of the Environment. *Plant Cell* 2 (April): 275–77.

Cubillos, Alberto, and Shigeru Suzuki. 1991. Japan Assists Chile in Launching INIA Genetic Resources Conservation Program. *Diversity* 7(1–2): 38–39.

Dahlberg, Kenneth A. 1979. *Beyond the Green Revolution: The Ecology and Politics of Global Agricultural Development*. New York: Plenum.

——— 1987. Redefining Development Priorities: Genetic Diversity and Agroecodevelopment. *Bulletin of Science, Technology, and Society* 7(3): 367–82.

Dattée, Yvette. 1986. La sélection conservatrice. In *La diversité des plantes légumières: Hier, aujourd'hui et demain*. Ed. Bureau des Ressources Génétiques, pp. 91–94. Paris: Bureau des Ressources Génétiques.

Davis, R. Hunt Jr. 1986. Agriculture, Food, and the Colonial Period. In *Food in Sub-Saharan Africa*. Ed. Art Hansen and Della E. McMillan, pp. 151–68. Boulder CO: Lynn Rienner.

Day, Boysie E. 1978. The Morality of Agronomy. In *Agronomy in Today's Society*. Ed. J. W. Pendleton, pp. 19–27. Special Publication 33. Madison WI: American Society of Agronomy.

De Bow, James D. B. 1854. *Statistical View of the United States, Being a Compendium of the Seventh Census*. Washington DC: A. O. P. Nicholson.

DeGeorge, Richard. 1979. The Environment, Rights, and Future Generations. In *Ethics and the Problems of the 21st Century*. Ed. K. Goodpaster and K. Sayre, pp. 93–105. Notre Dame: University of Notre Dame Press.

Diversity. 1987. FAO Commission Seeks Compromise on Undertaking. *Diversity* 11:15–16.

——— 1988a. ASTA Statement of Basic Concepts and Principles for Patent and PVP Rights for the Seed Industry. *Diversity* 14:27.

——— 1988b. New ARS Release Policy. *Diversity* 13:15.

——— 1989. Quandary Over Plant Patenting Brings Diverse Groups of Experts Together. *Diversity* 5(2–3): 35–37.

——— 1990. Uruguay Round of GATT Provides New Forum for Debating Germplasm Ownership Issues. *Diversity* 6(3–4): 39–40.

——— 1991. Report for U.S. Congress Forecasts Needs for National Genetic Resources Program. *Diversity* 7(4): 23–25.

——— 1992a. What Are They Saving: First Reactions to the Biodiversity Conference. *Diversity* 8(2): 9.

——— 1992b. Exclusion of CGIAR Germplasm Banks from Convention on Biological Diversity Raises Ire of Non-governmental Organizations. *Diversity* 8(3): 5.

——— 1992c. Clinton-Gore Administration: A New Era for Global Genetic Resources? *Diversity* 8(4): 23, 24.

——— 1992d. Expert Panels on Biodiversity Convention Convened by UNEP. *Diversity* 8(4): 6.

——— 1992e. After the Rio Earth Summit: Nations Move to Fulfill Commitments. *Diversity* 8(3): 4.

——— 1993a. International Conference on the Convention on Biological Diversity: National Interests and Global Imperatives. *Diversity* 9(1–2): 11.

——— 1993b. Global and Domestic Concerns Dominate Genetic Resources Advisory Council Meeting. *Diversity* 9(1–2): 61–62.

——— 1993c. President Clinton Establishes New White House Office on Environmental Policy. *Diversity* 9(1–2): 42.

——— 1993d. U.S. Plant Introduction Update. *Diversity* 9(1–2): 72.

Doyle, Jack. 1985a. *Ethical Aspects of Biotechnology*. Washington DC: Environmental Policy Institute.

——— 1985b. *Altered Harvest: Agriculture, Genetics, and the Future of the World's Food Supply.* New York: Viking Press.

Duesing, John. 1989. Plant Patenting as Seen by a Plant Breeding Professional. In *Patenting Life Forms in Europe, Proceedings of a Conference at the European Parliament,* pp. 27–30. Barcelona: International Coalition for Development Action.

Dugowson, Marc. 1986. Du cru au cuit . . . au froid. In *La planète alimentaire: Voyage au pays de l'agro-alimentaire*. Ed. Patrick Bernier, pp. 14–16. Paris: Editions de la Cité des Sciences et de l'Industrie.

Dunlap, Thomas R. 1981. *DDT: Scientists, Citizens, and Public Policy.* Princeton: Princeton University Press.

Duvick, Donald N. 1977. Major United States Crops in 1976. *Annals of the New York Academy of Sciences*. 287:86–96.

Duvick, Donald N., and William L. Brown. 1989. Plant Germplasm and the Economics of Agriculture. In *Biotic Diversity and Germplasm Preservation: Global Imperatives*. Ed. Lloyd Knutson and Allan K. Stoner, pp. 499–513. Dordrecht: Kluwer Academic Publishers.

Ecomusée. 1988. *Ecomusée du pays de Rennes: La ferme de la Bintinais, un espace agricole à découvrir.* Rennes: Ecomusée du Pays de Rennes.

Editora Abril. 1986. *Almanaque Abril '86*. São Paulo: Editora Abril.

Eisenberg, Rebecca S. 1987. Proprietary Rights and the Norms of Science in Biotechnology. *Yale Law Journal* 97 (December): 177–231.

Elster, Jon. 1983. *Explaining Technical Change*. Cambridge: Cambridge University Press.

Emery, James M. 1987. Patent Law: Obviousness, Secondary Considerations, and the Nexus Requirement. In *Annual Survey of American Law,* 1:117–41. New York: New York University.

Encina, Francisco A. 1955. *Nuestra inferioridad económica: sus causas, sus consecuencias*. Santiago, Chile: Editorial Universitaria.

Esquinas-Alcázar, José T. 1987. Plant Genetic Resources: A Base for Food Security. *Ceres* 20 (July–August): 39–45.

Evenson, Donald D., and Robert E. Evenson. 1983. Legal Systems and Private Sector Initiatives for the Invention of Agricultural Technology in Latin America. In *Technical Change and Social Conflict in Agriculture: Latin American Perspectives*. Ed. Martin Piñeiro and Eduardo Trigo, pp. 189–233. Boulder CO: Westview Press.

Ewan, Joseph. 1969. *A Short History of Botany in the United States*. New York: Hafner.

Fairchild, D. 1938. *The World Was My Garden: Travels of a Plant Explorer.* New York: Charles Scribner's Sons.

Farmer, Fannie Merritt. 1896. The Boston Cooking-School Cookbook. Boston: Little, Brown.

Fejer, S. O. 1966. The Problem of Plant Breeders' Rights. *Agricultural Science Review. Third Quarter:* 1–7.

Fishman, Ram. 1986. *The Handbook for Fruit Explorers*. Chapin IL: North American Fruit Explorers.

Fite, Gilbert C. 1966. *The Farmer's Frontier, 1865–1900*. New York: Holt, Rinehart and Winston.

Fitzgerald, Deborah. 1990. *The Business of Breeding: Hybrid Corn in Illinois, 1890–1940*. Ithaca: Cornell University Press.

Food and Agriculture Organization. 1985. *Country and International Institutions' Response to Resolution 8/83*. Rome: FAO, Commission on Plant Genetic Resources, CPGR/85/3.

——— 1986a. *Progress Report on the International Undertaking on Plant Genetic Resources*. Commission on Plant Genetic Resources, Second Session, Rome, 16–20 March 1987. CPGR/87/4, December 1986.

——— 1986b. *Study on Legal Arrangements with a View to the Possible Establishment of an International Network of Base Collections in Gene Banks Under the Auspices or Jurisdiction of FAO*. Commission on Plant Genetic Resources, Second Session, Rome, 16–20 March 1987. CPGR/87/6, December 1986.

——— 1987a. *Extract of the Twenty-Second Session of the FAO Conference. Resolution 8/83: International Undertaking on Plant Genetic Resources*. Commission on Plant Genetic Resources, Second Session, Rome, 16–20 March 1987. CPGR/87/Inf.3, January 1987.

——— 1987b. *Report of the Second Session of the Commission on Plant Genetic Resources*. Rome: FAO, Commission on Plant Genetic Resources, CPGR/87/REP.

——— 1987c. *Study of Legal Arrangements with a View to the Possible Establishment of an International Network of Base Collections in Gene Banks Under the Auspices or Jurisdiction of FAO*. Rome: FAO, Commission on Plant Genetic Resources, CPGR/87/6-Rev.

——— 1991a. *Report by the Chairman of the Working Group on Its Fourth Session*. Rome: FAO, Commission on Plant Genetic Resources, CPGR-91/3.

——— 1991b. *The Global System for the Conservation and Utilization of Plant Genetic Resources*. Rome: FAO, Commission on Plant Genetic Resources, CPGR/91/5.

——— 1991c. *Strategies for the Establishment of a Network of in Situ Conservation Areas*. Rome: FAO. Commission on Plant Genetic Resources, CPGR/91/6.

——— 1991d. *The Programme of IBPGR and FAO/IBPGR Cooperative Agreements*. Rome: FAO, Commission on Plant Genetic Resources, CPGR/91/11.

——— 1991e. *Second Progress Report on Legal Arrangements with a View to the Establishment of an International Network of Base Collections in Gene Banks Under the*

Auspices or Jurisdiction of FAO. Rome: FAO, Commission on Plant Genetic Resources, CPGR/91/13.

——— 1991f. *Memorandum of Understanding on Programme Cooperation Between the Food and Agriculture Organization of the United Nations and the International Board for Plant Genetic Resources*. Rome: FAO, Commission on Plant Genetic Resources, CPGR/91/Inf.4.

Ford-Lloyd, Brian, and Michael Jackson. 1984. Plant Gene Banks at Risk. *Nature*. 308:683.

Fottorino, Eric. 1989. La nostalgie de la France verte. *Le Monde* 46(13718): 1.

Fowler, Cary, Eva Lachkovics, Pat Roy Mooney, and Hope Shand. 1988. The Laws of Life: Another Development and the New Biotechnologies. *Development Dialogue*, nos. 1–2.

Fowler, Cary, and Pat Mooney. 1990. *Shattering: Food, Politics, and the Loss of Genetic Diversity*. Tucson: University of Arizona Press.

Frankel, Otto H. 1985. Genetic Resources: The Founding Years. Part One. *Diversity* 7:26–29.

——— 1986a. Genetic Resources: The Founding Years. Part Two. *Diversity* 8:30–32.

——— 1986b. Genetic Resources: The Founding Years. Part Three. *Diversity* 9:30–33.

——— 1988. Genetic Resources: Evolutionary and Social Responsibilities. In *Seeds and Sovereignty*. Ed. Jack R. Kloppenburg Jr., pp. 19–46. Durham: Duke University Press.

Friedland, William H., and Tim Kappel. 1979. *Production or Perish: Changing the Inequalities of Agricultural Research Priorities*. Santa Cruz: Project on Social Impact Assessment and Values, University of California.

Frischtak, Claudio R. 1990. The Protection of Intellectual Property Rights and Industrial Technology Development in Brazil. In *Intellectual Property Rights in Science, Technology, and Economic Performance*. Ed. Francis W. Rushing and Carole Ganz Brown, pp. 61–98. Boulder CO: Westview Press.

Fuller, Steve. 1988. *Social Epistemology*. Bloomington: Indiana University Press.

——— 1993. *Philosophy of Science and Its Discontents*. New York: Guilford Press.

Furtado, Maria Helena. 1988. Brazil Engineers a Biotech Revolution. *South*, no. 92 (June):91–93.

Gadamer, Hans-Georg. 1975. *Truth and Method*. New York: Seabury.

Gadbaw, R. Michael, and Rosemary E. Gwynn. 1988. Intellectual Property Rights in the New GATT Round. In *Intellectual Property Rights: Global Consensus, Global Conflict*. Ed. R. Michael Gadbaw and Timothy J. Richards, pp. 38–88. Boulder CO: Westview.

Gaillard, Jacques. 1991. *Scientists in the Third World*. Lexington: University Press of Kentucky.

Gaitskell, Arthur. 1959. *Gezira: A Story of Development in the Sudan*. London: Faber and Faber.

Galeano, Eduardo. 1973. *Open Veins of Latin America: Five Centuries of the Pillage of a Continent*. New York: Monthly Review Press.

Gallager, J. D. 1993. The United States Signs the Biodiversity Convention. *Diversity* 9(1–2): 40–41.

Gámez, Rodrigo. 1989. Threatened Habitats and Germplasm Conservation: A Central American Perspective. In *Biotic Diversity and Germplasm Preservation, Global Imperatives*. Ed. Lloyd Knutson and Allan K. Stoner, pp. 477–92. Dordrecht: Kluwer Academic Publishers.

General Accounting Office. 1981. *Better Collection and Maintenance Procedures Needed to Help Protect Agriculture's Germplasm Resources*. Washington DC: U.S. Government Printing Office.

Genetic Resources Action International. 1992. Why Farmer-Based Conservation and Improvement of Plant Genetic Resources? In *Growing Diversity: Genetic Resources and Local Food Security*. Ed. David Cooper, Renée Vellvé, and Henk Hobbelink, pp. 1–16. London: Intermediate Technology Publications.

George, Kathryn Paxton. 1988. Biotechnology and Biodiversity. *Journal of Agricultural Ethics* 1 (winter): 175–92.

Giacometti, Dalmo C. 1988. Introdução e intercâmbio de germoplasma. In *Encontro sobre recursos genéticos (anais)*. Ed. Samira Miguel Campos de Araujo and Juan Ayala Osuna, pp. 43–45. Meeting on Genetic Resources held in Jabotical, SP, 12–14 April 1988. Jabotical SP: FCAVJ/UNESP AND CENARGEN/EMBRAPA.

Giacometti, Dalmo C., and Clara O. Goedert. 1989. Brazil's National Genetic Resources and Biotechnology Center Preserves and Develops Valuable Germplasm. *Diversity* 5(4): 8–11.

Giedion, Siegfried. 1975. *Mechanization Takes Command*. New York: Norton.

Gilstrap, Margerite. 1961. The Greatest Service to Any Country. In *Seeds: Yearbook of Agriculture*, 78:18–26. Washington DC: USDA.

Goldstein, Daniel J. 1988. Molecular Biology and the Protection of Germplasm: A Matter of National Security. In *Seeds and Sovereignty*. Ed. Jack R. Kloppenburg Jr., pp. 315–37. Durham: Duke University Press.

——— 1989. Building Local Capability in Latin America. In *RIS Biotechnology Revolution and the Third World: Challenges and Policy Options*, pp. 385–419. New Delhi: Research and Information System for the Non-Aligned and Other Developing Countries (RISNODEC).

Góngora, Mario. 1960. *Origen de los "inquilinos" de Chile central*. Santiago: Universidad de Chile, Seminario de Historia Colonial.

Goodman, David, Bernado Sorj, and John Wilkinson. 1987. *From Farming to Biotechnology: A Theory of Agro-Industrial Development*. Oxford: Basil Blackwell.

Goodman, R. M. 1992. By Opposing Convention on Biological Diversity, U.S. Biotechnology Industry Missed an Opportunity to Play a Constructive Role. *Diversity* 8(3): 28–29.

Goonatilake, Susantha. 1982. *Crippled Minds: An Exploration into Colonial Culture*. New Delhi: Vikas Publishing House.

Gore, Albert. 1992. *Earth in the Balance*. New York: Houghton Mifflin.

Gottfried, Robert. 1992. On Gardening and Human Welfare. *Agriculture and Human Values* 9(4): 36–47.

Grall, Jacques, and Bertrand Lévy. 1985. *La guerre des semences*. Paris: Librairie Arthème Fayard.

Grynszpan, Flavio. 1990. Case Studies in Brazilian Intellectual Property Rights. In *Intellectual Property Rights in Science, Technology, and Economic Performance*. Ed. Francis W. Rushing and Carole Ganz Brown, pp. 99–112. Boulder CO: Westview.

Guillard, René. 1986. Point de vue d'un selectionneur au service d'un centre maraicher. In *La diversité des plantes légumières: Hier, aujourd'hui et demain*. Ed. Bureau des Ressources Génétiques, pp. 49–52. Paris: Bureau des Ressources Génétiques.

Guither, Harold. 1972. *Heritage of Plenty*. Danville IL: Interstate Publishers.

Hamilton, Alexander. 1958. *The Mind of Alexander Hamilton*. Edited by Saul K. Padover. New York: Harper.

Hanson, J. 1985. *Procedures for Handling Seeds in Genebanks*. Practical Manuals for Genebanks, no. 1. Rome: IBPGR Secretariat.

Hanson, J., J. T. Williams, and R. Freund. 1984. *Institutes Conserving Crop Germplasm: The IBPGR Global Network of Genebanks*. Rome: IBPGR Secretariat.

Hardon, J. J. 1989. Industrial Patents, Plant Breeding and Genetic Resources: A Plant Breeder's View. In *Patenting Life Forms in Europe*. Ed. International Coalition for Development Action, pp. 34–37. Brussels: International Coalition for Development Action.

Hargrove, T. R., V. L. Cabanilla, and W. R. Coffman. 1985. *Changes in Rice Breeding in 10 Asian Countries, 1965–84*. Research Paper III. Manila: International Rice Research Institute.

Hariot, Thomas. 1960. Food and Farming in Aboriginal Virginia. In *Readings in the History of American Agriculture*. Ed. Wayne D. Rasmussen, pp. 14–23. Urbana IL: University of Illinois Press.

Harlan, Jack R. 1975. Our Vanishing Genetic Resources. *Science* 188 (9 May): 618–21.

——— 1984. Gene Centers and Gene Utilization in American Agriculture. In *Plant Genetic Resources: A Conservation Imperative*. Ed. Christopher W. Yeatman, David Kafton, and Garrison Wilkes, pp. 111–29. Boulder CO: Westview.

——— 1988. Seeds and Sovereignty: An Epilogue. In *Seeds and Sovereignty*. Ed. Jack R. Kloppenburg Jr., pp. 356–62. Durham: Duke University Press.

Harlan, H. V., and M. L. Martini. 1936. Problems and Results of Barley Breeding. In *Yearbook of Agriculture*, pp. 303–46. Washington DC: U.S. Department of Agriculture.

Hart, Daniela. 1993. Academics in Brazil Criticize Plan to Revise Patent Law. *Chronicle of Higher Education*, 30 June: A33.

Haudricourt, André G., and Louis Hédrin. 1987. *L'homme et les plantes cultivées*. Paris: Editions A.-M. Métailié.

Hawkes, John G. 1985. *Plant Genetic Resources: The Impact of the International Agricultural Research Centers*. Washington DC: World Bank.

Hays, Samuel P. 1959. *Conservation and the Gospel of Efficiency*. Cambridge MA: Harvard University Press.

Heidegger, Martin. 1977. *The Question Concerning Technology and Other Essays*. New York: Harper & Row.

Hermitte, Marie-Angèle. 1988. Histoires juridiques extravagantes: La reproduction végétale. In *L'homme, la nature, et le droit*. Ed. Bernard Edelman and Marie Angèle-Hermitte, pp. 40–82. Paris: Christian Bourgois.

——— 1989. Patenting Life Forms: The Legal Environment. In *Patenting Life Forms in Europe, Proceedings of a Conference at the European Parliament*, pp. 15–16. Barcelona: International Coalition for Development Action.

Hermitte, Marie-Angèle, and Pierre-Benoit Joly. 1991. *Biotechnologies et brevets: Presentation des differents modèles juridiques et analyse de leur impact sur la dynamique de l'innovation dans les industries des semences*. Grenoble: Université des Sciences Sociales.

Hightower, Jim. 1973. *Hard Tomatoes, Hard Times*. Cambridge MA: Schenckman.

Hiss, Tony. 1989a. Encountering the Countryside: I. *New Yorker* 65 (21 August): 40–69.

——— 1989b. Encountering the Countryside: II. *New Yorker* 65 (28 August): 37–63.

Hobbelink, Henk. 1987a. *Biotechnology and Third World Agriculture: New Hope or False Promise?* Barcelona: International Coalition for Development Action.

——— 1987b. *Más allá de la revolución verde: Las nuevas tecnologías genéticas para la agricultura. ¿Desafío o desastre?* Barcelona: LERNA/ICDA.

Horwitz, Morton J. 1977. *The Transformation of American Law*. Cambridge MA: Harvard University Press.

Hoyt, Erich. 1988. Conserving the Wild Relatives of Crops. Rome: IBPGR, IUCN, and WWF.

Hymowitz, Theodore. 1984. Dorsett-Morse Soybean Collection Trip to East Asia: 50 Year Retrospective. *Economic Botany* 38:378–88.

IAC. 1987. *1887–1987 tem historias para contar*. Campinas, São Paulo: Instituto Agronômico de Campinas (IAC).

Instituto de Investigaciones Agropecuarias (INIA-CHILE). 1987. *Memoria anual*. Santiago: INIA.

International Board of Plant Genetic Resources. 1986. *El germoplasma vegetal en los paises del cono sur de América Latina*. Rome: Consejo Internacional de Recursos Fitogenéticos.

——— 1991. *Annual Report, 1990*. Rome: IBPGR.

——— 1992. *Annual Report, 1991*. Rome: IBPGR.

International Service for National Agricultural Research (ISNAR). 1987. *International Workshop on Agricultural Research Management: Report of a Workshop*. The Hague: ISNAR.

INRA. 1982? *Groupe d'Etude et de Contrôle des Variétés et des Semences*. Paris: INRA.

Jabs, Carolyn. 1984. *The Heirloom Gardener*. San Francisco: Sierra Club Books.

Jeffrey, Donald D., and Stanley D. Schlosser. 1990. The Plant Description Requirement of United States Patent Law. *Diversity* 6 (3–4): 42–43.

Jeremy, David John. 1977. Damming the Flood: British Government Efforts to Check the Outflow of Technicians and Machinery, 1740–1843. *Business History Review* 51 (Spring): 1–34.

Jobet, Julio César. 1955. *Ensayo critico del desarrollo económico social de Chile*. Santiago, Chile: Editorial Universitaria.

Jobim, Tom. 1988. *Jardim Botânico do Rio de Janeiro*. Rio de Janeiro: Expressão e Cultura.

Johnson, S. P. 1993. *The Earth Summit: The United Nations Conference on Environment and Development (UNCED)*. London: Graham and Trotman/Martinus Nijhoff.

Joly, Pierre-Bénoit. 1989. Should Seeds Be Patentable: Elements of an Economic Analysis. In *Patenting Life Forms in Europe: Proceedings of a Conference at the European Parliament*, pp. 17–21. Barcelona: International Coalition for Development Action.

Joly, Pierre-Benoit, and Michel Trommetter. 1991. World Regulation of Genetic Resources: Is the Model of Common Heritage Sustainable? Paper presented at the International Symposium of the Federation of Institutes of Advanced Studies, Nairobi.

Jonas, Hans. 1984. *The Imperative of Responsibility*. Chicago: University of Chicago Press.

——— 1985. Towards a Philosophy of Technology. In *Philosophy, Technology, and Human Affairs*. Ed. Larry Hickman, pp. 6–24. College Station TX: Ibis Press.

Juma, Calestous. 1989. *The Gene Hunters: Biotechnology and the Scramble for Seeds*. Princeton: Princeton University Press.

Justice, Oren L. 1961. The Science of Seed Testing. In *Seeds: Yearbook of Agriculture*, 78:407–13. Washington DC: USDA.

Kaan, F., and A. Boyat. 1988. Programme cooperatif français de gestion dynamique de la variabilité génétique chez le maïs. Unpublished document. Montpellier: INRA.

Kannenberg, Lyndon W. 1984. Utilization of Genetic Diversity in Crop Breeding. In *Plant Genetic Resources: A Conservation Imperative*. Ed. Christopher W. Yeatman, David Kafton, and Garrison Wilkes, pp. 93–109. Boulder CO: Westview Press.

Kaufman, C. S. 1992. Realizing the Potential of Grain Amaranth. *Food Reviews International* 8(1):5–21.

Kay, Cristobal. 1977. The Development of the Chilean Hacienda System, 1850–1953. In *Land and Labour in Latin America*. Ed. Kenneth Duncan and Ian Rutledge, pp. 103–39. Cambridge: Cambridge University Press.

Keystone Center. 1991. *Oslo Plenary Session. Final Consensus Report: Global Initiative for the Security and Sustainable Use of Plant Genetic Resources*. Keystone CO: Keystone Center.

Kingman, Sharon. 1989. Plant Patent Faces New Legal Challenge. *New Scientist* 124 (16 December): 10.

Kloppenburg, Jack R., Jr. 1988. *First the Seed: The Political Economy of Plant Biotechnology, 1492–2000*. New York: Cambridge University Press.

——— 1990. No Hunting! Scientific Poaching and Global Biodiversity. *Z Magazine* 3 (September): 104–8.

Kloppenburg, Jack R. Jr., and Daniel Lee Kleinman. 1986. The Common Bowl: Plant Genetic Interdependence in the World Economy. Paper presented at the American Association for the Advancement of Science Annual Meeting, Philadelphia.

——— 1988. Seeds of Controversy: National Property Versus Common Heritage. In *Seeds and Sovereignty*. Ed. Jack R. Kloppenburg Jr., pp. 173–203. Durham: Duke University Press.

Klose, Nelson. 1950. *America's Plant Heritage*. Ames: Iowa State College Press.

Knapp, S. A. 1910. The Farmer's Cooperative Demonstration Work. In *Yearbook of the United States Department of Agriculture, 1909*, 16:153–61. Washington DC: U.S. Government Printing Office.

Knutson, Lloyd, and Allan Stoner, eds. 1989. *Biotic Diversity and Germplasm Conservation*. Dordrecht: Kluwer Academic Publishers.

Krohn, Wolfgang, and Wolf Schafer. 1983. Agricultural Chemistry: The Origin and Structure of a Finalized Science. In *Finalization in Science: The Social Orientation of Scientific Progress*. Ed. Wolf Schafer, pp. 17–52. Dordrecht: D. Reidel.

Krosin, Kenneth E. 1985. Are Plants Patentable Under the Utility Patent Act? *Journal of the Patent and Trademark Office Society* 67 (May): 220–38.

Lacy, William B., Lawrence Busch, and Laura R. Lacy. 1991. Public Perceptions of Agricultural Biotechnology. In *Agricultural Biotechnology: Issues and Choices*. Ed. Bill R. Baumgardt and Marshall A. Martin, pp. 139–62. West Lafayette IN: Purdue University Agricultural Experiment Station.

Latour, Bruno. 1987. *Science in Action: How to Follow Scientists and Engineers Through Society.* Milton Keynes, England: Open University Press.

——— 1993. *We Have Never Been Modern*. Cambridge: Harvard University Press.

Leiss, William. 1972. *The Domination of Nature*. New York: Braziller.

Leonard, Thom. 1988. *The Grain Exchange: 1988 Spring Seed List*. Salina KN: Land Institute.

Levenstein, Harvey A. 1988. *Revolution at the Table*. New York: Oxford University Press.

Lévi-Strauss, Claude. 1969. *The Raw and the Cooked*. New York: Harper & Row.

Levin, Richard C., Alvin K. Klevorick, Richard R. Nelson, and Sidney G. Winter. 1987. Appropriating the Returns from Industrial Research and Development. *Brookings Papers on Economic Activity* 3:783–820.

Levine William, and A. Little. 1987. Biotech Spurs Most Competition. *Seed World* 125 (6): 19–20.

Levy, Jay. 1987. FAO Commission Seeks Compromise on Undertaking. *Diversity,* no 11:15–16.

Linck, Nancy J. 1985. Patentable Subject Matter Under Section 101 – Are Plants Included? *Journal of the Patent and Trademark Office Society* 67 (September): 489–506.

Lipschitz, Henri. 1986. Le contrôle du marché des semences et plantes des espèces légumières. In *La diversité des plantes légumières: Hier, aujourd'hui et demain*. Ed. Bureau des Ressources Génétiques, pp. 85–89. Paris: Bureau des Ressources Génétiques.

Lleras, Eduardo. 1988. Coleta de recursos genéticos vegetais. In *Encontro sobre recursos genéticos: Annals of the Meeting on Genetic Resources Held in Jaboticabal*. Ed. M. C de Araujo Samira and Juan A. Osuna, pp. 23–42. São Paulo, Brazil.

Locke, John. 1955. *Of Civil Government*. Chicago: Henry Regnery.

Lorain, John. 1960. Observations on Indian Corn and Potatoes. In *Readings in the History of American Agriculture*. Ed. Wayne D. Rasmussen, pp. 61–63. Urbana: University of Illinois Press.

——— 1975. Letter. In *Agriculture in the United States: A Documentary History*. Ed. Wayne D. Rasmussen, 1:416–26. New York: Random House.

Loveman, Brian. 1988. *Chile: Legacy of Hispanic Capitalism*. New York: Oxford University Press.

Lyman, A. B. 1960. Alfalfa Seed. In *Readings in the History of American Agriculture*. Ed. Wayne D. Rasmussen, pp. 95–102. Urbana: University of Illinois Press.

Macer, R. C. F. 1975. Plant Pathology in a Changing World. *Transactions of the British Mycological Society* 65 (December): 351–67.

Machiavelli, Niccolo. 1985. *The Prince*. Translated by Harvey C. Mansfield Jr. Chicago: University of Chicago Press.

Marchenay, Philippe. 1987. *A la recherche des variétés locales de plantes cultivées*. Paris: Muséum National d'Histoire Naturelle.

Marshall, D. R. 1989. Limitations to the Use of Germplasm Collections. In *The Use of Plant Genetic Resources*. Ed. A. H. D. Brown, O. H. Frankel, D. R. Marshall, and J. T. Williams, pp. 105–20. Cambridge: Cambridge University Press.

Marshall, D. R., and A. H. D. Brown. 1981. Wheat Genetic Resources. In *Wheat Science – Today and Tomorrow*. Ed. L. T. Evans and W. J. Peacock, pp. 21–40. Cambridge: Cambridge University Press.

Marshall, Eliot. 1991. The Patent Game: Raising the Ante. *Science* 252 (5 July): 20–24.

Michigan Crop Improvement Association. 1990. *The Plant Variety Improvement Act: Improvement Through Research*. East Lansing: Michigan Crop Improvement Association.

Mody, Ashoka. 1990. New International Environment for Intellectual Property Rights. In *Intellectual Property Rights in Science, Technology, and Economic Performance*. Ed. Francis W. Rushing and Carole Ganz Brown, pp. 203–39. Boulder CO: Westview Press.

Moffat, Anne Simon. 1991. Bumper Transgenic Plant Crop. *Science* 253 (5 July): 33.

Monaghan, Floyd V., and Alain F. Corcos. 1990. The Real Objective of Mendel's Paper. *Biology and Philosophy* 5:267–92.

Monroe, James. 1948. *Monroe's Defense of Jefferson and Freneau Against Hamilton*. Edited by Phillip M. Marsh. Oxford OH: Privately printed.

Montecinos, Camila, and Miguel A. Altieri. 1991. *Status and Trends in Grass-Roots Crop Genetic Conservation Efforts in Latin America*. Berkeley: Consorcio Latinoamericano sobre Agroecologia y Desarrollo.

Mooney, Pat Roy. 1983. The Law of the Seed: Another Development and Plant Genetic Resources. *Development Dialogue* 1–2:1–172.

——— 1985a. The Law of the Lamb. *Development Dialogue*, no. 1:103–8.

——— 1985b. The Law of the Seed Revisited: Seed Wars at the Circo Massino. *Development Dialogue*, no. 1:139–52.

——— 1986. O escândalo das sementes: O domínio na produção de alimentos. São Paulo: Livraria Nobel.

——— 1989. Biotechnology and North-South Conflict. In *RIS Biotechnology Revolution and the Third World: Challenges and Policy Options*, New Delhi: Research and

Information System for the Non-Aligned and Other Developing Countries (RISNODEC).

Moraga-Rojel, Jubel. 1989. Science, Technology, Socioeconomic, and Political Forces in the Shaping of Chilean Agricultural Research. Ph.D. dissertation, University of Kentucky.

——— 1990. Plant Germplasm and Agricultural Biotechnology: Prospects for Modern Agricultural Technology in Chile. *Revista Latinoamericana de las Ciencia y la Technologia – QUIPU* 7 (May-August): 217–31.

Mulkay, Michael. 1979. *Science and the Sociology of Knowledge*. London: George Allen & Unwin.

Murray, W. G. 1946. Struggle for Land Ownership. In *A Century of Farming in Iowa, 1846–1946*. Ed. Staff of the Iowa State College and the Iowa Agricultural Experiment Station, pp. 1–17. Ames: Iowa State College Press.

Myers, Norman. 1983. *A Wealth of Wild Species*. Boulder CO: Westview.

Nabhan, Gary. 1988. Southwest Project Cuts Across Cultural, National Boundaries. *Plant Conservation* 3(2): 1, 8.

Nabhan, Gary, and Kevin Dahl. 1987 [1985]. Role of Grassroots Activities in the Maintenance of Biological Diversity: Living Plant Collections of North American Genetic Resources. In *The 1987 Harvest Edition*. Ed. Kent Whealy, pp. 35–71. Decorah IA: Seed Savers Exchange.

National Academy of Sciences. 1975. *Underexploited Tropical Plants with Promising Economic Value*. Washington DC: NAS.

National Research Council, Board on Agriculture, Committee on Managing Global Genetic Resources. 1991. *Managing Global Genetic Resources: The US National Plant Germplasm System*. Washington DC: National Academy Press.

——— 1993. *Managing Global Genetic Resources: Agricultural Crop Issues and Policies*. Washington DC: National Academy Press.

Nichols, A. 1993. Interior Secretary Babbitt Takes Lead on National Biological Survey. *Diversity* 9(1–2): 42–44.

Norton, Bryan G. 1987. *Why Preserve Natural Diversity?* Princeton: Princeton University Press.

Nourse, Edwin. 1924. *American Agriculture and the European Market*. New York: McGraw-Hill.

Oasa, Edmund K. 1981. The International Rice Research Institute and the Green Revolution: A Case Study on the Politics of Agricultural Research. Ph.D. dissertation, University of Hawaii.

Office of Technology Assessment. 1986. *Intellectual Property Rights in an Age of Electronics and Information*. Washington DC: U.S. Government Printing Office.

——— 1987. *Technologies to Maintain Biological Diversity.* Washington DC: U.S. Government Printing Office.

Olivieri, Isabelle, and Jean-Marie Prosperi. 1989. La differenciation des populations et le rassemblement et la conservation des ressources génétiques. Paper presented at a workshop on the Genetics of Populations, La Londe-les-Maures.

Orton, Thomas J. 1988. New Technologies and the Enhancement of Plant Germplasm Diversity. In *Seeds and Sovereignty.* Ed. Jack R. Kloppenburg Jr., pp. 145–70. Durham: Duke University Press.

Parsons, F., C. Garrison, and K. Beeson. 1961. Seed Certification in the United States. In *Seeds, Yearbook of Agriculture,* 78:394–400. Washington DC: USDA.

Paschoal, Adilson Dias. 1986. Prefácio do tradutor. In *O escândalo das sementes: O domínio na produção de alimentos.* Ed. Pat Roy Mooney, pp. xiii–xxv. São Paulo: Livraria Nobel.

Pastore, José. 1978. Science and Technology in Brazilian Development. In *Science, Technology, and Economic Development.* Ed. William Beranek Jr. and Gustav Ranis, pp. 233–87. New York: Praeger.

Pastore, José, and Eliseu R. A. Alves. 1984. Reforming the Brazilian Agricultural Research System. In *Brazilian Agriculture and Agricultural Research.* Ed. Levon Yeganiantz, pp. 117–28. Brasilia, D.F.: Department of Diffusion of Technology/EMBRAPA.

Paul, Harry W. 1985. *From Knowledge to Power: The Rise of the Science Empire in France, 1860–1939.* Cambridge: Cambridge University Press.

Peacock, W. J. 1989. Molecular Biology and Genetic Resources. In *The Use of Plant Genetic Resources.* Ed. A. H. D. Brown, O. H. Frankel, D. R. Marshall, and J. T. Williams, pp. 363–76. Cambridge: Cambridge University Press.

Perkins, John H. 1990. The Rockefeller Foundation and the Green Revolution, 1941–1956. *Agriculture and Human Values* 7 (summer-fall): 6–18.

Pimentel, David, Ulrich Stachows, David A. Takacs, Hans W. Brubaker, Amy R. Dumas, John J. Meaney, John A. S. O'Neil, Douglas E. Onsi, and David B. Corzilius. 1992. Conserving Biological Diversity in Agricultural/Forestry Systems. *BioScience* 42 (May): 354–62.

Pinstrup-Andersen, Per. 1992. Food Security and Food Policy in a Changing World. Keynote address presented at the Eighth World Congress for Rural Sociology, University Park PA.

Plucknett, Donald L., N. J. Smith, J. T Williams, and N. M. Anishetty. 1983. Crop Germplasm Conservation and Developing Countries. *Science* 220:163–69.

Plucknett, Donald, Nigel J. H. Smith, J. T. Williams, and N. Murthi Anishetty. 1987. *Gene Banks and the World's Food.* Princeton: Princeton University Press.

Porter, Donna Viola, and Robert O. Earl. 1990. *Nutrition Labeling: Issues and Directions for the 1990s.* Washington DC: National Academy Press.

Porter, G. 1992. The United States and the Global Biodiversity Convention: The Case for Participation. Washington DC: Environmental and Energy Study Institute Paper Series on Environment and Development no. 1.

Prest, John M. 1981. *The Garden of Eden: The Botanic Garden and the Re-creation of Paradise*. New Haven: Yale University Press.

Radnitzky, Gerard. 1973. *Contemporary Schools of Metascience*. 3d ed. Chicago: Henry Regnery.

Raeburn, P. 1992. The Convention on Biological Diversity: Landmark Earth Summit Pact Opens Uncertain New Era for Use and Exchange of Genetic Resources. *Diversity* 8(2): 4–7.

Raemond, R. D. 1992. Transformation of IBPGR into New Institute. *Diversity* 8(1): 4.

Raines, Lisa J. 1991–92. Protecting Biotechnology's Pioneers. *Issues in Science and Technology*. (winter): 33–39.

Rasmussen, Wayne D. 1974. *Agriculture in the United States: A Documentary History*. 4 vols. New York: Random House.

——— 1960. *Readings in the History of American Agriculture*. Urbana: University of Illinois Press.

Ravetz, Jerome. 1971. *Scientific Knowledge and Its Social Problems*. New York: Oxford University Press.

Rawls, John. 1971. *A Theory of Justice*. Cambridge MA: Belknap Press of Harvard University Press.

Raymond, Thierry de. 1985. La question des semences: Point de vue. In *A travers champs: Agronomes et géographes*, pp. 57–99. Bondy: ORSTOM.

Reboul, Claude. 1977. Determinants sociaux de la fertilité des sols. *Actes de la Recherche en Sciences Sociales* 17–18 (November): 85–112.

Reich, Leonard S. 1985. *The Making of American Industrial Research: Science and Business at GE and Bell, 1876–1926*. Cambridge: Cambridge University Press.

Reichert, Walt. 1982. Agriculture's Diminishing Diversity. *Environment* 24 (November): 6–11, 39–43.

Research and Information System (RIS). 1989. *Biotechnology Revolution and the Third World: Challenges and Policy Options*. New Delhi: Research and Information System for the Non-Aligned and Other Developing Countries.

Richards, Timothy J. 1988. Brazil. In *Intellectual Property Rights: Global Consensus, Global Conflict*. Ed. R. Michael Gadbaw and Timothy J. Richards, pp. 149–85. Boulder CO: Westview.

Riehl, S. 1993. Biodiversity Issues Rank High on Congressional Environmental Agenda. *Diversity* 9(1–2): 53–55.

Rindos, David. 1980. Symbiosis, Instability, and the Origins and Spread of Agriculture: A New Model. *Current Anthropology* 21 (December): 751–65.

Rivard, Paul E. 1974. Textile Experiments in Rhode Island, 1788–1789. *Rhode Island History* 33 (May): 35–45

Rives, Max. 1980. Annexe I: Plantes Cultivées. In Conservation des ressources Génétiques. Ed. B. Vissac and R. Cassini. Paris, unpublished report presented to the minister of agriculture.

Rogoff, Martin H., and Stephen L. Rawlins. 1987. Food Security: A Technological Alternative. *BioScience* 37(11): 800–807.

Rouse, Joseph. 1987. *Knowledge and Power*. Ithaca: Cornell University Press.

Rousseaux, Luc. 1987. *La quatrième ressource: La patrimoine génétique végétale*. Toulouse: Pioneer France Maïs.

Rozek, Richard P. 1990. Protection of Intellectual Property Rights. In *Intellectual Property Rights in Science, Technology, and Economic Performance*. Ed. Francis W. Rushing and Carole Ganz Brown, pp. 31–46. Boulder CO: Westview.

Rural Advancement Fund International (RAFI). 1986. *Los bancos de semillas locales. Un material de apoyo*. Manitoba: Centro de Education y Tecnologia.

Sagoff, Mark. 1988. *The Economy of the Earth: Philosophy, Law and the Environment*. Cambridge: Cambridge University Press.

Saint-Albin, Michel de. 1986. L'assiette en 1000 morceaux. In *La planète alimentaire: Voyage au pays de l'agro-alimentaire*. Ed. Patrick Bernier, pp. 17–22. Paris: Editions de la Cité des Sciences et de l'Industrie.

Schilling, H. 1982. The New Seed Monopolies. *Raw Material Report* (Sweden) 1, no. 3:40–51.

Schmid, A. Allan. 1989. *Cost-Benefit Analysis*. Boulder CO: Westview.

Schwartzman, Simon. 1978. Struggling to Be Born: The Scientific Community in Brazil. *Minerva* 16(4): 545–80.

Scott, Roy Vernon. 1970. *The Reluctant Farmer: The Rise of Agricultural Extension to 1914*. Urbana: University of Illinois Press.

Sedjo, Roger A. 1988. Property Rights and the Protection of Plant Genetic Resources. In *Seeds and Sovereignty*. Ed. Jack R. Kloppenburg Jr., pp. 294–314. Durham: Duke University Press.

Sen, Amartya, and Bernard Williams. 1982. *Utilitarianism and Beyond*. Cambridge: Cambridge University Press.

Sepúlveda, Sergio. 1959. El trigo Chileno en el mercado mundial. Santiago, Chile: Editorial Universitaria.

Serpette, Raoul. 1986. La certification des semences et plantes des espèces légumières et leur commercialisation. In *La diversité des plantes légumières: Hier, aujourd'hui et demain*. Ed. Bureau des Ressources Génétiques, pp. 75–84. Paris: Bureau des Ressources Génétiques.

Shands, Henry. 1992. Who Owns Germplasm and the Implications for Genebanks. Paper presented at the annual meeting of the Agricultural Research Institute, Washington DC.

——— 1994. Some Potential Impacts of the United Nations Environment Program's Convention on Biological Diversity on the International System of Exchanges of Food Crop Germplasm. In *Conservation of Plant Genetic Resources and the UN Convention on Biological Diversity*. Ed. Daniel Witmeyer and Michael S. Strauss, pp. 27–38. Washington DC: American Association for the Advancement of Science.

Shrivastava, R. K., et al. 1984. Rice Germplasm Conservation and Evaluation Activities at J. N. Agricultural University campus, Raipur, MP, India. *International Rice Research Newsletter* 9(1): 7–8.

Silva, George H. 1985. In Vitro Storage of Potato Tuber Explants and Subsequent Plant Regeneration. *HortScience* 20:139–40.

Smith, Nigel J. H. 1985. *Botanic Gardens and Germplasm Conservation*. Honolulu: University of Hawaii Press.

SOLAGRAL. 1988. *Les semences au coeur des mutations agricoles*. Paris: Solagral.

Souza Silva, José de. 1989. Biotechnology in Brazil and Prospects for South-South Cooperation. In *Biotechnology Revolution and the Third World: Challenges and Policy Options*. Ed. RIS, pp. 420–42. New Delhi: Research and Information System for the Non-Aligned and Other Developing Countries (RISNODEC).

Straus, Joseph. 1987. The Relationship Between Plant Variety Protection and Patent Protection for Biotechnological Inventions from an International Viewpoint. *International Review of Industrial Property and Copyright Law* 18(6): 723–37.

Strauss, Anselm. 1978. *Negotiations: Varieties, Contexts, Processes, and Social Order*. San Francisco: Jossey-Bass.

Strauss, D. G. 1993. Spirit of Cooperation Overshadows Disharmonies as FAO Marks Tenth Anniversary. *Diversity* 9(1–2): 4–6.

Stuckey, Ronald L. 1978. *Development of Botany in Selected Regions of North America Before 1900*. New York: Arno Press.

Taylor, Christopher T., and Z. A. Silberston. 1973. *The Economic Impact of the Patent System: A Study of the British Experience*. Cambridge: Cambridge University Press.

Tigchelaar, Edward C. 1986. Tomato Breeding. In *Breeding Vegetable Crops*. Ed. Mark J. Bassett, pp. 135–71. Westport CT: AVI.

Towill, Leigh E., and Eric E. Roos. 1989. Techniques for Preserving Plant Germplasm. In *Biotic Diversity and Germplasm Preservation: Global Imperatives*. Ed. Lloyd Knutson and Allan K. Stoner, pp. 379–403. Dordrecht: Kluwer Academic Publishers.

Treat, Payson Jackson. 1910. *The National Land System, 1785–1820*. New York: E. B. Treat.

Tucker, Barbara. 1984. *Samuel Slater and the Origins of the American Textile Industry, 1790–1860*. Ithaca: Cornell University Press.

Turner, F. 1988. A Field Guide to the Synthetic Landscape. *Harper's Magazine* 276 (April): 49–55.

United States Department of Agriculture. 1984. Position Paper of the United States Department of Agriculture on the International Plant Germplasm System. Washington DC: USDA. Photocopy.

——— 1993. *Inventory of Agricultural Research, Fiscal Year 1992*. Washington DC: USDA, Cooperative State Research Service.

Universidad Austral de Chile and International Board for Plant Genetic Resources (UACH-IBPGR). 1984. *Recursos fitogenéticos*. Valdivia, Chile: Anales del Simposio realizado en la Universidad Austral de Chile.

UPOV. 1991. *Final Draft. International Convention for the Protection of New Varieties of Plants of December 2, 1961, as Revised at Geneva on November 10, 1972, on October 23, 1978, and on March 19, 1991*. Geneva: International Union for the Protection of New Varieties of Plants.

Usdin, S. 1992. Biotech Industry Played Key Role in US Refusal to Sign Bioconvention. *Diversity* 8(2): 8–9.

Venezian, Eduardo. 1987. *Chile and the CGIAR Centers: A Study of Their Collaboration in Agricultural Research*. Study Paper 20. Washington DC: World Bank.

Virgil. 1982. *The Georgics*. Translated by L. P. Wilkinson. New York: Penguin Books.

Vissac, B., and R. Cassini. 1980. *Conservation des ressources génétiques*. Paris, unpublished report presented to the minister of agriculture.

Wallerstein, Immanuel. 1974. *The Modern World-System: Capitalist Agriculture and the Origins of the European World-Economy in the Sixteenth Century*. New York: Academic Press.

Watson, Elkanah. 1819. *History of the Rise, Progress, and Existing State of the Berkshire Agricultural Society*. Albany NY: E. & E. Hasford.

Weiss, Martin G., and Elbert L. Little Jr. 1961. Variety Is a Key Word. In *Seed: Yearbook of Agriculture*, 78:359–64. Washington DC: USDA.

Whealy, Kent. 1988a. *The 1988 Winter Yearbook*. Decorah IA: Seed Savers Exchange.

——— 1985. *The Garden Seed Inventory: Inventory of Seed Catalogs – Listing All Non-Hybrid Vegetable and Garden Seeds Still Available in the United States and Canada*. Decorah IA: Seed Savers Publications.

——— 1988b. *The Garden Seed Inventory: Inventory of Seed Catalogs – Listing All Non-Hybrid Vegetable and Garden Seeds Still Available in the United States and Canada*. 2d ed. Decorah, IA: Seed Savers Publications.

——— 1992. *The Garden Seed Inventory: Inventory of Seed Catalogs – Listing All Non-Hybrid Vegetable and Garden Seeds Still Available in the United States and Canada.* 3d ed. Decorah IA: Seed Savers Publications.

Whealy, Kent, and Arllys Adelmann, eds. 1986. *Seed Savers Exchange: The First Ten Years.* Decorah IA: Seed Savers Publications.

Wilkes, Garrison. 1983. Current Status of Crop Plant Germplasm. *Critical Reviews in the Plant Sciences.* 1:133–81.

——— 1984. Germplasm Conservation Toward the Year 2000. In *Plant Genetic Resources: A Conservation Imperative.* Ed. Christopher W. Yeatman, David Kafton, and Garrison Wilkes, pp. 131–64. Boulder CO: Westview.

——— 1991. *In Situ* Conservation in Agricultural Systems. In *Biodiversity: Culture, Conservation, and Ecodevelopment.* Ed. Margery L. Oldfield and Janis B. Alcorn, pp. 86–101. Boulder CO: Westview.

Williams, J. T. 1989. Practical Considerations Relevant to Effective Evaluation. In *The Use of Plant Genetic Resources.* Ed. A. H. D. Brown, O. H. Frankel, D. R. Marshall, and J. T. Williams, pp. 235–44.Cambridge: Cambridge University Press.

Withers, L. A. 1989. In Vitro Conservation and Germplasm Utilisation. In *The Use of Plant Genetic Resources.* Ed. A. H. D. Brown, O. H. Frankel, D. R. Marshall, and J. T. Williams, pp. 309–34. Cambridge: Cambridge University Press.

Witt, Steven C. 1985. *Biotechnology and Genetic Diversity.* San Francisco: California Agricultural Land Project (CALP).

Wood, David. 1988. Crop Germplasm: Common Heritage or Farmers' Heritage? In *Seeds and Sovereignty.* Ed. Jack R. Kloppenburg, pp. 274–89. Durham: Duke University Press.

Wright, Thomas C. 1982. *Landowners and Reform in Chile: The Sociedad Nacional de Agricultura, 1919–1940.* Urbana: University of Illinois Press.

Zeitlin, Maurice. 1984. *The Civil Wars in Chile (or the Bourgeois Revolutions That Never Were).* Princeton: Princeton University Press.

Index

IN THE SUSTAINABLE FUTURE SERIES

Volume 1
Ogallala: Water for a Dry Land
John Opie

Volume 2
Building Soils for Better Crops: Organic Matter Management
Fred Magdoff

Volume 3
Agricultural Research Alternatives
William Lockeretz and Molly D. Anderson

Volume 4
Crop Improvement for Sustainable Agriculture
Edited by M. Brett Callaway and Charles A. Francis

Volume 5
Future Harvest: Pesticide-Free Farming
Jim Bender

Volume 6
A Conspiracy of Optimism: Management of the National Forests since World War Two
Paul W. Hirt

Volume 7
Green Plans: Greenprint for Sustainability
Huey D. Johnson

Volume 8
Making Nature, Shaping Culture: Plant Biodiversity in Global Context
Lawrence Busch, William B. Lacy, Jeffrey Burkhardt, Douglas Hemken, Jubel Moraga-Rojel, Timothy Koponen, and José de Souza Silva